Can Star Systems be Explored?

The Physics of Star Probes

Can Star Systems be Explored?

The Physics of Star Probes

Lawrence B. Crowell

Alpha Institute of Advanced Studies, USA

World Scientific

NEW JERSEY · LONDON · SINGAPORE · BEIJING · SHANGHAI · HONG KONG · TAIPEI · CHENNAI

Published by

World Scientific Publishing Co. Pte. Ltd.

5 Toh Tuck Link, Singapore 596224

USA office: 27 Warren Street, Suite 401-402, Hackensack, NJ 07601

UK office: 57 Shelton Street, Covent Garden, London WC2H 9HE

British Library Cataloguing-in-Publication Data
A catalogue record for this book is available from the British Library.

CAN STAR SYSTEMS BE EXPLORED?
The Physics of Star Probes

ISBN-13 978-981-270-617-1
ISBN-10 981-270-617-8
ISBN-13 978-981-270-618-8 (pbk)
ISBN-10 981-270-618-6 (pbk)

Printed in Singapore.

To Junko

Preface

Many students are motivated by applications of physical theory instead of presentations of pure theory. Many students gain an interest in relativity early on from exposure to science fiction or from speculations on space travel. The depiction of futuristic space flight and exotic methods of spaceflight are often a catalyst for young people's interest in physics. This is the spirit in which this book was written. The intention here is to present Newton's mechanics, relativity and astrophysics with the question of whether it is possible to send a probe to another star. There have been some presentations on this prospect, such as the Daedalus project and Robert Forward's Starwisp. The prospect is tantalizing, and something which might take fruition in the future. In this light, the prospect of interstellar probes is discussed in the light of known physics. The reader is given the option to determine whether this is at all possible.

The current age presents us with some apparent limits on scientific knowledge and technological progress. This has been seen with the increased technical and financial problems which accompany the building of ever larger particle accelerators. It is also seen in the shortfall with the reality of manned spaceflight when compared to futuristic expectations of past decades. In spite of these apparent limits, scientific research has continued remarkably well. While manned space exploration appears blunted compared to early expectations, space science has made huge advances in astronomical and planetary research. Possibly, particle physics will develop new methods that circumvent current limitations, or even find new ways to convert matter to energy. Similar advances may occur with the exploration of space with probes. Probes have been sent to all of the planets in the solar system, except for Kuiper belt planets, and to a number of asteroids and comets. These probes have velocities limited by current technology, but with ion propulsion and VASMIR systems, it is certainly possible that larger craft will be sent into the solar system at much higher velocities. It is possible that technical limitations will be overcome and that in due time, probes might be sent to some of the nearby stars.

Physics has penetrated considerably into the public sphere. There has been a recent increase in popular interest in quantum mechanics and relativity. Popularizations of an increasingly sophisticated level have made their way into the general reading populace. Indeed, the distinction between texts, research books and popularizations has begun to fade in recent years. This book is meant to exist within the boundary between a text and a popularization. The general reader may find this an interesting route to reading about mechanics and relativity, and students may find this to be an entertaining supplemental source in addition to classroom lectures and texts. It might also stimulate some interest in the physics and engineering of space probes and in preliminary studies on the technical requirements for future star probes.

L. B. Crowell

Contents

Preface vii

1. Exploration of Star Systems 1

2. Newtonian Mechanics 5

3. The Physics of Rocketry and Spaceflight 17

4. Power Systems for Spaceflight 31

5. Elements of Astrodynamics 43

6. Special Relativity 49

7. The Relativistic Rocket 67

8. The Photon Sail 79

9. Scientific and Technical Requirements 91

10. Electromagnetically Accelerated Nano-bots 101

11. Exotic Propulsion Methods 107

12. The Interstellar Neighborhood 119

13. Will Humans go to the Stars? 129

14. Solar System Stability and the Likelihood of Earth-like Planets 141

15. Life on Earth and in the Universe 163

16. Appendix 183

Bibliography 185

Index 189

Chapter 1

Exploration of Star Systems

Is space exploration of other star systems possible? The idea dominates much of science fiction. The old *Star Trek* television series popularized the idea of interstellar flight during the years leading up to the Apollo space missions to the moon. Ten years later there came the *Star Wars* movie, which brought an upsurge in public interest with space travel after a nadir in its popularity following the cancelled Apollo program. There followed a spate of science fiction movies involved with star travel, where most of them were of poor quality. Star travel persists in our imaginations, but the currently prospects appear very remote. The only space missions that might be considered as star travel are the Pioneer and Voyager probes which have left the solar system. They will never send any data back concerning planetary systems around other stars, where it will also take tens of thousands of years for them to reach distances comparable to those between stars. However, this does not mean that star travel is impossible.

Spacecraft are sent into the solar system at velocities 15–20 km/sec. This speed, while fast by most considerations, is far smaller than the speed of light $c = 299,997$ km/sec. It consequently takes one of our spacecraft about 15–20,000 years to travel a light year. This is due to limits of our current state of rocketry. Chemical rockets can only send a craft to a maximum velocity of around 30 km/sec. The Voyager spacecrafts had to use a gravitational slingshot approach, where they absorbed some angular momentum and energy as they flew past Jupiter, in order to exit the solar system. Nuclear propulsion will do better, rating at around 100 km/sec, but this again falls very short of reaching velocities approaching the speed of light. So our space systems are far too sluggish to reach other stars.

By measuring the Doppler shift of stars extrasolar planets have been detected, and a few imaged. Most of these are gas giant planets similar to Jupiter. Some extrasolar systems have been found to have several gas giant planets arrayed in various orbits. There is a rich variety of stellar system structure that has been discovered. Optical interferometers may soon image these planets with considerable detail. They may further image smaller terrestrial planets similar to Venus or Earth. We may explore other star systems in the near future by such means from near Earth space. However, with some 145 extrasolar system identified so far with planets, and potentially thousands waiting to be found, there is the prospect that a planet might be found with optical signatures similar to Earth. If a planet of this sort is detected it will confirm Carl Sagan's thesis that life exists elsewhere in the universe. However, there is no way that such remote detection will ever reveal to us the nature of life on this planet. A scientific premium would be placed upon sending a spacecraft to this star system to examine this life in detail.

The major thesis of this book concerns sending space probes to other stars. Exuberant ideas of large interstellar spacecraft with people on board are not likely to obtain, at least not any time in the foreseeable future. The physics and technology required to send an un-piloted probe to another star at some significant fraction of light speed are formidable challenges in of themselves. It has to be honestly admitted that manned spaceflight is also very expensive and tends not to produce the same measure of real results obtained with space probes. Unfortunately it appears that other planets in our solar system offer little for us humans, who are these squishy watery complexes of biomass. The same may well be the case for other planetary systems. So schemes of colonizing other planets are at best problematic. Further, an extrasolar planet identified as biologically active might be a serious biohazard to any human being stepping forth onto its surface. In fact we may find this to be the case for the planet Mars. In addition this risks contaminating Mars with earthly bacteria. So it is a certainty that the first of spacecrafts to reach other stars will be un-piloted robotic craft, where it may be unlikely that humans will ever directly travel to another star except by the sort of virtual reality seen from our probes sent to planets in our solar system.

Many science fiction portrayals of star travel invoke warp drives and other schemes designed to short circuit the limitation of light speed. The classical laws of gravitation turn out to predict such spacetime structures, such as wormholes and warp drives. However, mass-energy couples to spacetime in general relativity and gravity very weakly. It requires the accumulation of large amounts of mass in order to create an appreciable gravity field. By the same reasoning it is very difficult to imagine how a system can be readily devised that is able to engineer spacetime curvatures. There are also problems with these exotic types of spacetime solutions as well. They violate the energy conditions of general relativity established by Hawking and Penrose. Consequently quantum fields which act as the source for spacetime curvature suffer serious pathologies. This will be discussed further on. It is unlikely that starships will ever be constructed with warp drive capability.

The preliminary answer to the opening question of this introduction, is yes. However, this yes has qualifications, as it is an affirmative answer to a possibility. If interstellar probes are sent into space it will likely be at least a half century or more from today. The duration for these missions will be measured in decades as well. Much can happen here on Earth in the mean time. Such stellar exploration can only happen if we manage to tackle a fair number of problems we face. These include, energy and resource depletion, global public health and pandemics, nuclear war, population pressures and the long term prospect of a global ecological collapse. This is a major qualifier, for stellar exploration depends upon a stable global situation lasting for at least another century. It further depends upon either some breakthroughs in our understanding of quantum field theory, or a future ability to construct large systems in space. In the first case this means that large amounts of anti-matter needs to be generated, or some way that conservation laws violated to convert mass directly to energy. If this is physically realistic a photon propelled relativistic rocket is possible. In the second this is required to construct large solar collectors and photon sail ships. As seen further on this requires construction in space on a very large scale. Substantive hurdles have to be overcome for either of these approaches to work.

So with this tentative affirmation I will proceed to illustrate the basic problems here. If such stellar exploration is never conducted at least I intend for this book to be a basic overview of basic mechanics and special relativity. These matters are here discussed in an entertaining format that differs from most books on elementary physics.

Chapter 2

Newtonian Mechanics

Isaac Newton first pondered space flight in a physically correct manner, but with the wrong technology. He illustrated orbital motion by considering a cannon on a high mountain or perch which shot a ball at a high enough velocity to keep missing the curvature of the Earth. Of course he meant this as a method of illustration, yet at the core he demonstrated how a satellite could be put in Earth orbit. By extension Jules Verne in his novel *From Earth to Moon*, described a trip to the moon in a large projectile shot from a cannon. Again the cannon was the method of propulsion. The obvious problem here that the crew of this spaceship would be flattened to splatters and bits due to the rapid acceleration to 11 km/sec. The first serious suggestion of rocket flight to space was made by Tsiolkovsky, who also derived the celebrated rocket equation. In the early 20th century Oberth began to lay out the engineering requirements for space rockets. Robert Goddard flew the first successful chemical rocket in 1926. Soon various investigators, von Braun, Winkler, Goddard, Korolev were building larger liquid propelled rockets. It took World War II for the rocket to come into its own. In 1944 Germany used the V-2 rocket to bomb England and other allied positions in a desperate attempt to reverse their diminishing fortunes of war. This was the first clear demonstration that a craft could reach the edge of the Earth's atmosphere by rocketry as the V-2 reached an altitude of 100 km and a speed of over 1000 m/sec.

With the end of the war rocket development in the United States and the Soviet Union proceeded into what then was called the space race. Of course it also lead to the missile race and the existence of Inter-Continental Ballistic Missiles that can deliver a cluster of nuclear bombs to a target and kill millions. Yet pick up a cell phone or make a credit card transaction and one is employing satellites. We get images from the surface of the Saturnian moon Titan, pictures of distant galaxies from the Hubble Space Telescope, watch the martian scenery from robots driving around the martian surface and get our daily weather reports. The rocket makes all of this possible.

To understand how rockets work it is important to understand the physics behind them. This means starting with basic Newtonian mechanics. Before Newton were Copernicus, Kepler and Galileo who respectively laid down the heliocentric solar system, a mathematical description planetary motion, and a description of motion seen on Earth. Galileo furnished some basic equations for the trajectory of an object under an acceleration. This development was stimulated by the introduction of gun powder into Europe, where previous ideas about the flight of a body were found to be inadequate. Gun powder is expensive after all, and a good physics for ballistic flight was needed. Yet even with these early 17th century accomplishments it still remained an open question as to the reason for these principles, and whether the motion of planets had anything to do with the motion of a body in flight here on Earth. Isaac Newton provided the answer in his three laws of motion and his inverse squared law of gravity. It was a grand synthesis of physics that unified the dynamics of bodies here on Earth with the motion of planets. Isaac Newton's *Philosophiae Naturalis Principia Mathematica* was published in 1687, where after that date it can be said that our views about the nature of the world changed forever [2.1].

Nature and Nature's laws lay hid in night; God said, Let Newton be! And all was light.
Alexander Pope

The laws of motion were laid down in the third book of the *Principia* called *De Mundi Systemate (On the system of the world)* which give the whole of dynamics. These are codified in the three laws of motion.

Newton's first law of motion is:

- A body in a state of motion will remain in that state of motion unless acted upon by a force.

This first law introduces the idea of a force by considering its absence. The first law defines an inertial reference frame. On this frame there are no external forces acting upon it and one will observe a body at rest there remaining at rest. Further, this observer may be on another inertial reference with some relative motion to this body, and this observer will see this body move with a constant velocity and remain in this state of linear motion. This first law illustrates something about the nature of inertia as laid down by Galileo in his equation for the linear motion of a body without acceleration $x = vt$, and constant velocity. It further gives the appropriate reference frame from which the other two laws of physics apply. In other words the laws of mechanics are correct for an observer in a reference frame that is not accelerating.

Newton's second law of motion is:

- The acceleration of a body, or equivalently the change in it momentum, is directly proportional to a force applied to it.

This is the dynamical principle of classical mechanics. The momentum of a body is its mass times its velocity $p = mv$, *mass times velocity*, where that velocity is determined by an observer in an inertial reference frame. The time rate of change of the momentum is given by calculus as $dp/dt = ma$, where the acceleration is $a = dv/dt$. The first law of motion tells us that the only reliable reference frame from which to observe these dynamics is from an inertial reference frame. From this inertial reference frame the acceleration a is measured. In order for this acceleration to exist there must be a force F exerted on the mass. This leads to the famous expression for the second law of motion

$$F = ma\,. \tag{2.1}$$

The unit of force $kg - m/sec^2$ is called a Newton N. The first law tells us that only the acceleration is directly measured and not the force. This is even the case for the measurement of a weight. A spring has a force $F = -kx$ where x is the extension of the spring and k is a constant. This equation indicates a spring exerts a force in the opposite direction of its extension, where this force and distance are related by the spring constant k with units of N/m. The measurement of a weight involves the measurement of $F = -mg$ for $g = 9.8m/s^2$ the acceleration of gravity here on the Earth's surface. The second law result $F = -kx = -mg$ infers the weight by the extension of a spring, which indirectly measures the force.

Newton's third law of mechanics is:

- Whenever one body exerts a force on a second body, the second body exerts a force of equal magnitude on the first in the opposite direction.

The third law tells us that ultimately momentum is conserved. The action of a force by the second law means that the material body exerting this force will experience an oppositely directed force of the same magnitude. So while there might be an $F = ma$ on one body there is ultimately an $-F = -m'a'$ exerted on another body or mass. The net force is $F - F = 0$ which conforms to the first law of motion. Further, if one body experiences a change in momentum δp it does so by inducing an opposite change in momentum $-\delta p$

on another mass. Fundamentally the third law tells us that the second law operates in a way that is homogeneous and isotropic. In other words how forces acts between bodies is independent of their position and orientation in space. The two forces are always in the opposite direction, and further that the second law of motion is independent of where in space this takes place. This means that Newtonian dynamics is independent of the translation of an inertial frame to another position or its rotational orientation. The first law also indicates that the third law is obeyed for any two masses who's center of mass is travelling at any velocity with respect to an inertial frame. Formally this means that space is isotropic and homogeneous and obeys a set of symmetries given by Galilean translations and orthogonal rotations.

Newton's second law tells us that mechanics is deterministic. Newton's second law of motion is a second order differential equation $F = md^2x/dt^2$, which remains the same if the time variables t is replaced by $-t$. This means that if the dynamics of a body is played backwards in time the dynamics conform to the same principle. Information concerning the initial configuration or state of a system is preserved by dynamics. Newton's laws then imply that the motion of everything can in principle be understood with arbitrary precision and for any time into the future or the past. The second order form of the differential equation for Newton's second law indicates a time reverse invariance. This later was found to imply conservation of energy. Newtonian mechanics was considered in the 18th century to illustrate the clockwork universe. However, it turns out these differential equations for systems involving three or more bodies, or masses, defy integration in a closed form. Classical mechanics is strictly deterministic, but that determinism in most cases is not exactly computable. This is particularly if there are three or more bodies interacting with each other. Isaac Newton in his computations ran into difficulties with modelling the solar system, where he suggested that God might have to intervene to keep it stable [2.2].

Newtonian mechanics unified the mechanics of bodies in motion terrestrially with the motion of the planets. Ultimately this meant that the solar system and beyond lost some of their attributed divine qualities. The motion of Mars is fundamentally no different from the motion of a cannon ball. Prior to Newton there was a general sense that some type of distance rule between bodies governed their motion. Kepler envisioned an inverse linear law for motion, where he lacked the three laws of mechanics and thought the rule should only apply in the plane of planetary motion. Newton's success was in demonstrating that this law of motion. As discussed below the connection with Kepler's law demonstrates that the law of gravity is an inverse square law.

King Oscar II of Sweden, in honor of his 60th birthday, offered a prize for a mathematical demonstration on the stability of the solar system. A year later in 1888 Henri Poincaré offered an alternative solution, by demonstrating that this question was not posed properly [2.3]. A system with three or more bodies is not generally integrable, and further any small perturbation on the system will have amplified effects on orbits. This means the system will radically diverge from an expected result at some future time by even the smallest of perturbation. The example in the stability of the solar system is the orbit of a satellite with a small mass in the reduced two-body problem for the two larger masses. It is easy to reduce the dynamics of two mutually interacting bodies to one body. Let this two-body problem be subjected to a perturbation by some distant or small mass. The perturbed two-body orbit is found to be close to an elliptical Keplerian orbit, or the orbit is an approximation to a Keplerian orbit, which is a solution to the reduced two-body problem. Poincaré considered a plane transverse to the Kepler orbit for the two-body problem, called a Poincaré section. When the actual orbit passes through this plane with each revolution it defines a point on the plane. The unperturbed two-body orbit always pass through the same point. The perturbed two-body orbit might under the right conditions pass through a repeated cycle of points, but in general the orbit hits a different point in this plane with every revolution. To an observer the developing array of these pass through points has no apparent pattern to their occurrence. This pattern in general has no discernable structure and appears completely random.

This was a precursor to the theory of chaos, which in this case is called deterministic chaos. It is a curious matter that a subject centered around $F = ma$, which is completely deterministic, gives rise to chaos. The term deterministic chaos indicates that in principle the dynamics can be integrated numerically arbitrarily into the future, but the computer would require an infinite floating point capability. Even the smallest truncation of a numerically integration of a three body problem will evolve into a significant contribution that results in a divergence in the actual motion from the numerical simulation. Yet, nature is deterministic and "knows" where it is heading, but we are not able to have that knowledge.

Isaac Newton was a co-founder of calculus, which he laid down in the first book of his *Principia De Motu Corporum (On the motion of bodies)*, along with Gottfried Leibniz, which Newton developed in a proximal way in order to work his laws of mechanics. His approach involved what he called "fluxions," which are infinitesimals. Leibniz introduce the differential, which has a more mathematically well defined concept by limits. Newton's fluxions had some of the content of differentials, but lacked precision and were largely less workable.

It is common to hear that Newtonian mechanics is somehow passé as having been replaced by relativity theory and quantum mechanics so that Newton's classical mechanics is approximate to situations of low velocities and a large scale. While this is true, I urge the reader to again read these three law of motion. It should be apparent that these are truly remarkable principles! There is the central dynamical principle in the second law of motion, which has great utility, but that this law obtains for a certain system of observers on inertial frames within a space that has a mathematical structure of symmetry of translations and rotations. Symmetry structures in more modern physics reveal a crucial connection: conservation principles (such as Newton's third law of motion) is equivalent to a principle of symmetry for the space of consideration. Newton's third law implies conservation of momentum with a symmetry to space. Emmy Noether was the first to clearly codify this connection between symmetry and a conservation law, where her insight was endorsed by Albert Einstein. Isaac Newton was the first to write dynamical principles that made this sort of connection.

The dynamics of $F = ma$ may be used to solve a wide range of problems. A small illustration might be in order here. The spring force is $F = -kx$, which says that a spring exerts a force proportional to its distention x. This force is equated by the second law to the mass times acceleration of the mass. So Newton's second law indicates that a mass on a spring will in general obey motion according to

$$F = mdv/dt = -kx\,. \tag{2.2}$$

Yet the velocity is the time rate of change of position or $v = dx/dt$. This leads to the differential equation

$$md^2x/dt^2 = -kx\,. \tag{2.3}$$

Without any explicit reference to the theory of differential equations, it is known that functions which satisfy this differential equation are trigonometric functions of sines and cosines. The position of the mass held by this spring is $x = Asin(\sqrt{k/m}\ t) + Bcos(\sqrt{k/m}\ t)$. The amplitudes A and B depend upon the initial conditions of the problem. Anyone familiar with trigonometric functions knows that this is oscillatory. The spring vibrates up and down in general if it is extended and then released.

For a pendulum in motion the situation is similar. If the angle of swing is small its motion approximates that of a mass on a spring. This results in a dynamical motion that depends only upon the length of the string holding the mass. The mass of the line is regarded as far less than the mass of the mass at the end, and is treated as irrelevant. This was something noted by Galileo, but he was unable to codify in terms of some consistent dynamics. This is a simple problem common in elementary physics texts.

Most elementary classes in physics treat the environment here on Earth as an inertial reference frame. However, this is strictly not the case since the Earth is rotating. However, these effects are small enough to be ignored. Further, local gravity $F = -mg$ for a body sitting on a surface is countered by a normal force N due to the strength of the material under the surface. Hence the net acceleration $a = (N/m - g)$ is zero. A body sitting on a surface is then said to be on an inertial reference frame. Yet this turns out not to be strictly the case for reasons due to general relativity. Yet for basic Newtonian mechanics, which is admitted to be an approximation, these assumptions are made with a good measure of safety.

Isaac Newton introduced another ingredient to his dynamics. This is the universal law of gravitation. Near the surface of the Earth there is a force of gravity $F = mg$, but this will be different for a region removed from the Earth's surface. Newton realized gravity operated between masses in general. Further, this attractive force is what binds planets in the solar system in their orbits. Newton derived a general rule for gravity. The story of Newton's apple, likely apocryphal, illustrates how he had the insight that when the apple fell to the Earth the apple also pulled the Earth to it, though very slightly. This is consistent with his third law of motion. So for two masses m_1 and m_2 separated by a distance r this force must be $F = F(m_1,\ m_2,\ r)$. This force must also fit within his second law of motion $F = ma$.

It is necessary to find the acceleration for a particle in circular motion. A circularly moving particle that defines some $\delta\phi$ in some time δt traverses an arc of length $r\delta\phi$. If at one point on the arc it has a velocity $\mathbf{v}$ tangent to the circle, then some $\delta\mathbf{v}$ is needed to change the velocity to that tangent to the second point $\mathbf{v} \rightarrow \mathbf{v} + \delta\mathbf{v}$. The change in velocity $\delta\mathbf{v}$ points towards the center of the circle. Figure 2.1 illustrates the existence of two similar triangles, one for the small pie slice in the orbital circle and the other for the velocity and its change. By the similarities of these two triangles it is

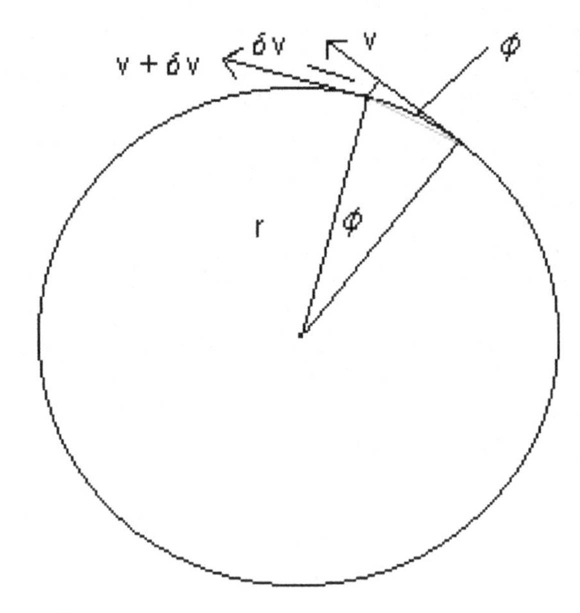

Fig. 2.1. The simple geometric construction for why the centripetal acceleration is v^2/r.

evident that

$$\frac{\delta v}{v} = \delta\phi \,, \tag{2.4}$$

with $v = rd\phi/dt$. Since $\delta v = dv/dt \times \delta t$ and $\delta\phi = d\phi/dt \times \delta t$ this may be written with some algebra as

$$\frac{dv}{dt} = \frac{v^2}{r} \,, \tag{2.5}$$

which is the centripetal acceleration required to keep a particle in circular motion. This may also be written for the angular velocity $\omega = d\phi/dt$ as

$$a = \frac{dv}{dt} = \omega^2 r \,, \tag{2.6}$$

and the second law of motion gives $F = mv^2/r$. This centripetal force points radially inward and is required to keep a mass in circular motion.

Newton then applied his famous $F = ma$ for the force of gravity written as $F = Gm_1m_2r^n$, where the exponent n is to be determined and G is the universal constant of gravitation. If the first mass m_1 is far larger than the second m_2 then the motion of interest is with this second mass. Newton's second law of motion is then

$$m_2\omega^2 r = Gm_1m_2r^n\,, \tag{2.7}$$

which gives the equation for the angular velocity

$$\omega^2 = Gm_2r^{n-1}\,. \tag{2.8}$$

For $n = -2$ this recovers Kepler's third law (1618): The square of the sidereal period, obtained from ω, of an orbiting planet is directly proportional to the cube of the orbit's semi-major axis. Newton's universal law of gravitation is then $F = -GMm/r^2$, where the negative sign indicates an inward radial direction for the force. In terms of vectors the gravity force is written in more generality as

$$\mathbf{F} = \frac{GMm}{r^3}\mathbf{r}\,, \tag{2.9}$$

for $r = \sqrt{\mathbf{r}\cdot\mathbf{r}}$ and $\cdot$ meaning the dot product of the radial vector with itself.

In 1664, Newton demonstrated how a body moving in a Keplerian ellipse experienced acceleration in the direction of one of the foci and inversely proportional, in magnitude, to the square of the distance from that focus. His second law $F = ma$ with Kepler's law gives the inverse square law. This demonstration of the inverse square principal from Keplerian ellipses is the converse of the modern day "textbook" illustration. Newton illustrated how motion in a Keplerian ellipse derives an inverse square law, whereas the standard exposition these days is to show that an inverse square law implies a Keplerian ellipse.

Newton did little with his observation for nearly two decades until implored by Robert Hooke, around 1680, to reconsider the question of how the inverse square rule implies motion in a Keplerian ellipse. Hooke drew upon a picture well known to many at this juncture, a picture wherein force "radiates" from a primary mass and therefore must fall off proportional to the inverse square of distance. Newton was reluctant to take this on again. Most upheld the conventional Cartesian vortex model, and Newton was not quick to abandon the idea. He was also troubled by the "action at a distance" implications of the inverse square law. Hooke confessed to Newton that he lacked the mathematical techniques required to handle the problem, though he had successfully investigated some other power laws for a central force. In later time Newton and Hooke became bitter enemies, largely due to Newton's unfriendly behavior.

This exchange led ultimately to the famous visit of Edmund Halley and Christopher Wren to Cambridge, where a dialog between them and Newton ensued. When asked by the pair what the motion of a body would be under a force with a magnitude proportional to the inverse square of the distance as of the form, Newton stated it would move in an ellipse. When asked how he knew Newton replied "Why, I have calculated it!"Newton's statement is probably true, but the calculation was probably a more fitful undertaking than Newton wanted to admit. In later years he famously claimed that his law of gravity was derived in 1664, but this is probably an enormous exaggeration. In any case, the immediate result was the publication of "De Motu,"and the immense "Principia," undertaken at Halley's urging.

Later in the 1690's Johann Bernoulli pointed out, with a measure of glee and pleasure, that neither in "De Motu" nor in the "Principia" did Newton actually derive Kepler's laws of motion from the inverse square law. Bernoulli was an ally of Leibniz in the calculus priority dispute and an enemy of Newton's, and thus eager to raise this objection. Rather, just as in the early work, he had shown the converse. Newton's answer was that the inverse square law gives the differential equation for a dynamical system. Solutions to differential equations exist and are unique. Hence Kepler's laws of motion then constitute a proof of the inverse square law of gravity.

These were Newton's great contribution of the theory of dynamics, where it might be said Newton invented physics. It should also be noted that Newton made great contributions to the subject of optics as well. This view of mechanics persisted for 218 years until Einstein introduced the special theory of relativity. During that two centuries considerable progress was made in advancing Newtonian mechanics. Joseph Louis Lagrange illustrated how Newton's laws could be reduced to a minimization principle. William Rowan Hamilton illustrated later how this could further be transformed into a system where position and momentum variables were treated equivalently. With Newtonian gravitation considerable progress was made in the calculation of planetary orbits, with considerable astronomical work to measure the orbital parameters of the planets. Mathematical techniques were developed to compute the perturbing influence of the various planets upon each other. Poincaré found that Newtonian mechanics was only exactly solvable for two mutually interacting bodies. For more interacting bodies certain methods of approximation must be applied. In particular if one out of three masses is small, or far removed from the other two, its influence can be treated with perturbation methods. Poincaré demonstrated that the stability of the solar system could not be computed in an exact solution of closed form.

Newton of course faced his detractors in his day. This was particularly for his law of gravitation. It suggests that a masses have invisible threads that connect each other. Forces are given by vectors along these threads. People raised objections to this, for this sounded almost magical. In the middle ages is was thought that angels pushed the planets around in their orbits, where it might be said that these threads appear to be just like angels, but where they push on a planet radially according to a prescription. These threads often go by the term action at a distance. However, it must be recalled that in Newtonian mechanics the force is not directly measured. The force is only inferred by an acceleration induced on a body. These threads are then model systems used to compute the observed effect. Newton's detractors were asking the wrong question. It has been said that this was "resolved" with Einstein's theory of general relativity. However, relativity simply replaces these threads of gravity with curvatures of spacetime. So this question about Newtonian gravity is transferred to the question, "What are space, time and spacetime?" Again it can only be said that spacetime is simply what it is defined to be in order to give the appropriate results which pertain to our observations.

This point here is worth discussing some, for it is a source of serious confusion. This confusion sometimes occurs with physicists. This has been seen with the hidden variablists, who have a hard time with some of some strange consequences of quantum mechanics. Hidden variables are proposed as a way to avoid the strange consequences of quantum mechanics. Yet in the 75 years since quantum mechanics was established in it early complete form there have been utterly no indications of hidden variables. It is most likely that these people are simply asking the wrong question. They are asking for something that physics simply can't deliver, such as asking what are the threads in Newtonian gravitation. This is because such questions are of an existential nature that is beyond the power of any science to determine. To ask whether the threads of gravity exist in of themselves is a meaningless question. Do these threads exist? We can't know, nor is the question at all useful. These threads are only useful as constructions which permit us to perform calculations. The hidden variablists attempt to impose an ontology, or an added level of being, to the nature of the quantum wave. Quantum mechanics appears unable to accommodate this, and experiments designed to test for hidden variables have constantly been null. Is there a deeper level of ontological structure to quantum waves?

The question appears to have no answer.

For anyone contemplating studies or a career in physics it is best to be wary of this. The non-question, or the unspeakable, in physics is an intellectual trap. Don't do this to yourself. Recognize these traps before you fall into them, and should you fall recognize that and get out!

There is often a bit of a debate as which is harder, mathematics or physics. To which it has to be pointed out that mathematics of a serious nature has a much older history. Euclid's *Elements* are not entirely easy to read. Ptolemy's *Almagest*, which is the geocentric model of the solar system, makes for a very dense mathematical reading, as it is based on Euclid. Yet physics was largely a mystery at the time these were written. Galileo and Kepler opened up the first cracks to the mystery, which Newton finally pried open. Physics can be said to have started in 1687, while western mathematics has its start with Pythagoras around 600 B.C.E.. Physics is dirtier in some ways than mathematics, in that physics has "friction." Aristotle said that for a body to be in motion required a constant force. Experience tells us this. We need to use fuel to keep our car going at a constant velocity, we need to keep pushing something along to keep moving it and so forth. Yet the muddling factor here is friction. Without friction Newton's laws are clearly seen. Friction is added as a force to Newton's laws so the everyday world can be seen in the Newtonian frame. Friction, or other issues, often cloud the essential problem at hand. This is the crucial difficulty in physics. Mathematics does not have "friction."

Chapter 3

The Physics of Rocketry and Spaceflight

Probably most readers are aware of Newton's third law as the basis for rocketry. The recoil of a gun is familiar to many, particularly those in the United States. If one throws a mass in one direction with a certain velocity there is a corresponding impulse that pushes you back. A rocket burns chemicals in some reaction, where the hot gasses produced are shot out the back nozzle. This change in momentum induced on the gas is compensated for by the opposite momentum change in the rocket body, plus the fuel it will have to burn to accelerate further. This was more or less known since the time of Newton, but was not developed well. Rockets were generally used for displays or by militaries as ways to lob explosives into a fort, which is referenced in Francis Scott Key's the rocket's red glare now used in the American national anthem. Even still the mortar and cannon were the preferred method of choice for bombardments. So for the most part rockets were generally thought of as toys.

Konstantin Tsiolkovsy was the first person to seriously ponder the use of rockets as a method for reaching beyond the Earth in the early 20th century. He outlined his advocacy of rocketry in his *The Exploration of Cosmic Space by Means of Reaction Devices.* He was also the first to apply physics to the issue of rocket flight, which at the time meant Newton's laws of motion. To obtain his rocket equation suppose there exists a mass M that fragments into $M - \delta M$ and δM, where $\delta M \ll M$. Further, suppose that δM flies away at a velocity V. The large mass will then experience a change in velocity $-\delta v$ so that

$$0 = \delta MV - (M - \delta M)\delta v\,. \tag{3.1}$$

If these increments are small, the term $\delta M \delta v$ may be ignored. This is the conservation of momentum for a brief increment in the rocket flight. From here in the calculus limit these increments become infinitesimal

$$0 = VdM - Mdv, \rightarrow \int_0^v dv' = V\int_M^m \frac{dM'}{M'}. \tag{3.2}$$

This leads to a final velocity of the rocket with mass m after expending a mass $M-m$ of burned fuel plus oxidant $v = Vln(M/m)$. For those familiar with logarithms should sense that this is a terrible result! It's not terrible for being wrong, its terrible in its implications. If the initial rocket mass is ten times that of the final mass the velocity of that terminal part will be 2.3 times the velocity of the exhaust plume. This multiplication by 2 to 3 is a standard economic multiplication factor. For a rocket with a plume velocity of 3500 m/sec and a final payload mass of .1 the initial mass the final velocity will be 8050 m/sec, which will place the craft into Earth orbit. This is the primary reason that it takes a rocket some 100 feet tall to put a much smaller spacecraft into space. This becomes all the more the case for a round trip vehicle such as it the Saturn-Apollo ship, which started off at 365 foot tall and around 3000 tons, and what came back was a small capsule and its crew of three.

Essentially this result means a rocket has to carry its propellant up with it. Most of the energy and reaction mass, the rocket plume, expended is done so in order the carry the fuel along the way. Yet there is no escape from this reality. There is some talk of a space elevator, a colossal structure that reaches out into geosynchronous orbit, that will get around this problem. However, I have serious doubts about the feasibility of this. Engineers must then try to get the plume velocity as fast as possible. The measure of this is the specific impulse, which is the plume velocity divided by the Earth's gravity at the surface $g = 9.8$ m/sec^2, $s = V/g$ [3.1]. Liquid chemical rockets have an upper limit of $s = 500 seconds$, and the space shuttle engines have $s = 459$ sec, which is close to the theoretical limit. Another physical quantity of use is thrust, which is the force of a rocket. This is $T = Vdm/dt = sg\ dm/dt$, where dm/dt is the rate that reaction mass is thrown out the nozzle of the rocket [3.1]. One strategy to reduce the impact of the rocket equation is to drop various rocket stages to eliminate lofting unneeded mass along the way. Obviously another is to reduce the mass of the rocket as much as possible. This constraint is a brutal limitation imposed on spaceflight.

Other methods of propulsion have been proposed. An ion rocket accelerates ions as the reaction mass. This can have a specific impulse of 3000 seconds. An ion propulsion rocket may be powered by either solar

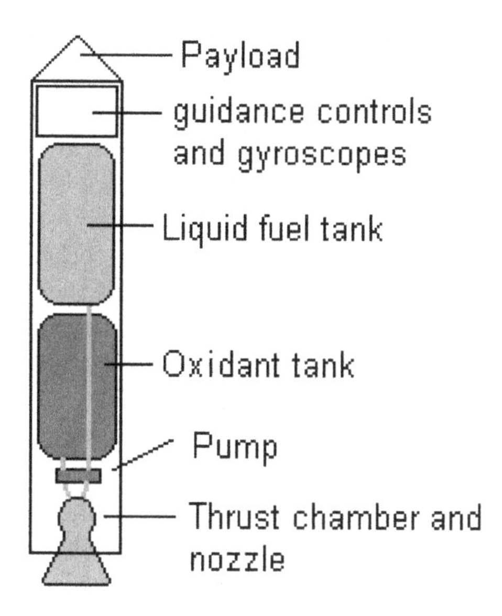

Fig. 3.1. Schematic for the basic liquid chemical rocket.

photovoltaic panels or a nuclear reactor. This works well for propulsion of a craft already in space, but as its thrust is low ion propulsion will not work for lifting off the Earth. A nuclear propulsion system, where hydrogen runs through heat pipes in a nuclear reactor may have a similar specific impulse. This might be useful for launch vehicles. Yet concerns over accidents have posed a serious barrier to its implementation. Another propulsion method is the Variable Specific Impulse Magnetoplasma Rocket (VASIMR), which employs electromagnetic radiation to impart energy to hydrogen, ionize it and magnetic fields then guide the plasma out the rocket at very high velocity. This may theoretically have a specific impulse of up to 30,000 sec and higher thrust than the ion rocket. The VASIMR is the highest specific impulse propulsion system being seriously considered. Yet this is insufficient for sending a spacecraft to the stars. What will be primarily addressed is the photon rocket. The photon rocket employs some means, such as matter-antimatter reactions, to convert much of its initial mass into photons. Since the plume velocity is the speed of light $c \simeq 3 \times 10^8$ m/sec the specific impulse is $s \simeq 3. \times 10^7$ sec, which is the upper limit of specific impulse.

The basic features of each of these propulsion methods are considered. The simplest and oldest form is the solid rocket. Essentially this is a can with some rocket fuel in a solid form packed within and an exit port for the burned gases. The simplest fuel is gunpowder, which forms the basis for firework display rockets and model rockets. Another form is the sulphur-zinc rocket, which can be easily made, but great care is advised as this can be explosively dangerous. Various improvements in chemistry and configurations of the propellant have resulted in higher specific impulses, but the upper limit is $s \simeq 200$ sec. Conversely solid propellant rockets do have high thrusts. They are mainly used for ballistic missiles, as they can be stored ready for launch over long time durations. Curiously they found spaceflight use with the space shuttle. The early pioneers of rocketry recognized that a superior method was required for effective space flight.

The liquid chemical rocket was first devised by Robert Goddard. It is comparatively simple in principle. It burns a fuel and an oxidant in a bottle with an exit nozzle, the thrust chamber. The fuel and oxidant are pumped into the thrust chamber by various means from separate tanks. A simple way of providing this feed is to pressurize the tanks. Other systems involve high pressure pumps which are powered by gasses channelled through pipes from the thrust chamber which drive a turbine which in turn powers a pump. This is remarkably simple, of course simple in principle. The complexities come with high temperatures and pressures in the thrust chamber. A crack or flaw can result in the rocket exploding, which early films indicate was a serious problem. To reduce heat damage often the oxidant is liquid oxygen, which is circulated around the thrust chamber before being injected in. Obviously other issues have to be addressed, such as oxidant and propellant vapors entering the rocket engine area causing explosions. The system for delivering fuel and oxidant into the thrust chamber needs to work at high pressure, where leaks could be explosive. Yet the chemical rocket is fairly reliable, and today their launches are comparatively routine.

Yet it is clear that the chemical rocket is marginally capable of lofting interplanetary spacecraft. Time of travel is measured in years. Further, chemical rockets are not able to loft very large spacecraft deep into interplanetary space. Chemical propulsion may well remain the method of choice for getting payloads off Earth and into space for some time. Yet once the craft is sent to escape velocity, $v \simeq 11$ km/sec, it would be convenient to have a method of propulsion that can send a craft to 100 km/sec or even 1000 km/sec to cut mission times proportionately. This requires alternative forms of propulsion, such as the ion rocket, nuclear propulsion, or VASIMR. For interstellar rocketry the photon rocket will be discussed in detail in Chapter 7.

The ion rocket is the most developed of these three [3.2]. An ion thruster uses beams of ions for propulsion. There are various methods for accelerating the ions. This system employs the strong electromagnetic force to accelerate ions to very high velocities. Since the mass of the ions is small this results in a high specific impulse $s \simeq 3000$ sec. Ion thrusters with a plume of high velocity ions have high specific impulse and a small amount of reaction mass constituting the plume. The small dm/dt means that the trust is very small. The power requirements for a high velocity ion plume is large when compared to chemical rockets. Since the thrust is low ion rockets can only be employed in space and are not effective for a launch vehicle.

The simplest ion rocket is essentially no different than a vacuum tube or a particle accelerator. Atoms are stripped of their electrons by electromagnetic means. The ions are then attracted to a highly negatively charged grid. There may be a series of these, where quadrupole magnets maintain ion beam stability. At the end of the process the electrons stripped from the ions are sent into the beam to recombine with the ions and the high velocity beam of ions, or recombined atoms, defines the rocket plume. This form of the ion drive is the electrostatic ion thruster.

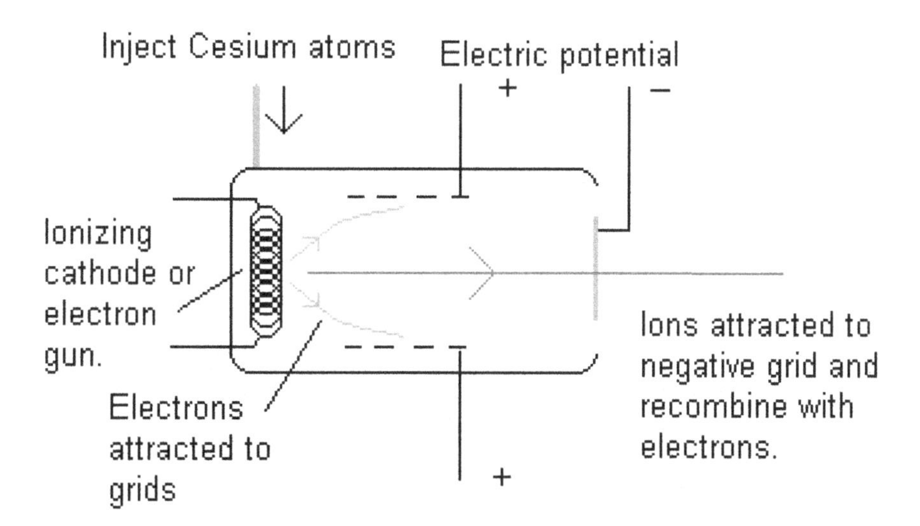

Fig. 3.2. Schematic for an ion propulsion unit.

The typical thrust of an ion rocket is a fraction of a Newton of force. The accelerations of an ion rocket is 10^{-3} "gees," $g = 9.8$ m/sec^2, or an acceleration around a centimeter per second squared. It is clear that this can't be used as a launch vehicle off the Earth's surface, it wouldn't lift off the ground. However, the exhaust plume travels at around 30 km/sec, where based on a 2.5 multiplication factor means the spacecraft may reach a final velocity of around 75 km/sec. Yet it is apparent that by $v = at$ that it takes 7,500,000 sec or a quarter of a year for the craft to reach this terminal velocity. This can reduce a four year trip to Jupiter to a single year, with similar time reductions for any interplanetary exploration.

Ion thrusters have been used by the Russians for maintaining orbits, called station keeping, such as with geosynchronous satellites. The solar wind can perturb the orbit of such a satellite and the ion rocket is used to correct the orbit. NASA developed the electrostatic xenon ion engine called NSTAR for use in their interplanetary missions. The successful space probe Deep Space 1 employed this thruster. Deep Space 1 is a prototype of an interplanetary spacecraft propelled by an ion motor. Hughes Aerospace has developed the XIPS (Xenon Ion Propulsion System) for performing station keeping on geosynchronous satellites.

The Russians developed a Hall effect ion rocket motor. This employs the Hall effect, where electrons are caught in a circular orbit within a magnetic field, to trap electrons and use them to strip the electrons off of atoms, which are then electromagnetically ejected out for thrust. This system has been used for station keeping of satellites.

The High Power Electric Propulsion, or HiPEP was ground tested in 2003 by NASA [3.3]. The HiPEP engine produces xenon ions by a combination of microwave and magnetic fields to oscillate electrons in the propellant atoms, causing the energetic electrons to escape free of the propellant atoms, to produce positive ions. This Electron Cyclotron Resonance (ECR) is similar to the VASIMR process.

These are high impulse, low thrust engines that have comparatively high power requirements. Ion thrust is likely to be the next step for interplanetary spacecraft. This will require either solar power or nuclear power to drive them, where the latter power system choice will most likely be employed for spacecraft targeted to the gas giant planets due to low solar illumination in these more outer regions of the solar system. For the same reason the RTG type decay reactors were used on the Galileo and Cassini spacecrafts. To explore the outer reaches of the solar system nuclear power systems are an obvious necessity. This means that safety considerations for their deployment from Earth are critical issues.

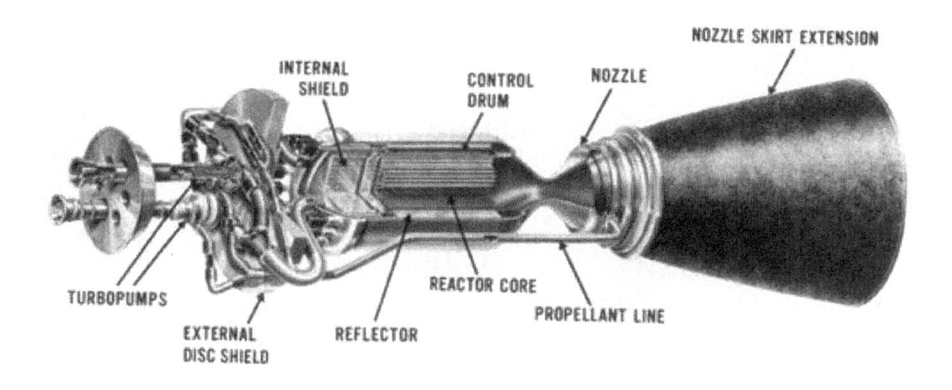

Fig. 3.3. Proposed design for the NERVA rocket engine.

The next form of propulsion beyond chemical rockets are nuclear propulsion units. The nuclear propelled rocket has some problematic safety and environmental issues, which have to be addressed. The essential idea is that liquid hydrogen is forced through heat pipes that run through a high temperature nuclear reactor [3.2]. The heat of the reactor is transferred to the hydrogen, which expands rapidly as the gas passes through the heat pipes to exit the engine out a nozzle at high velocity. The design of the NERVA rocket involved a solid core reactor. The solid-core can only be run at temperatures below the melting point of the materials used in the reactor core. This required lower temperature reduces thrust. A rocket engine is more efficient for high velocities of the plume, which means higher temperatures. A solid-core reactor design must be constructed of materials which remain strong at high temperatures. These material science limitations, where advanced materials melt at temperatures below temperatures which could be generated in a nuclear reaction, results in a loss of possible energy these reactions generate. Further, the system can't be run too close to the melting point, for otherwise while the core structure might not melt it will deform. An addition problem is that neutrons in the reactor core will induce damage to the crystalline structure of these materials, which will lower their tolerance to high temperatures. Generally the solid-core design is expected to deliver specific impulses on the order of 1000 sec, or about twice that of the space shuttle engines. However, the thrust of these rockets is comparatively low. The acceleration expected at this point is less than one gee, which means that nuclear propulsion on the engineering theory level is so far incapable of being a launch vehicle.

A crude form of this engine was actually built and tested in 1959, called Kiwi, where the designation indicated it was a static test system. This was essentially a hot running nuclear reactor of uranium oxide where liquid hydrogen was poured over the hot reactor. The system generated 70 megawatts (MW) and produced an plume at a temperature $2683K$. This lead to the Phoebus series of nuclear propulsion static test devices. This involved a much larger nuclear reactor. The final test series in this program was conducted in 1968. The static test engine was run for over 12 minutes at 4,000 MW. This is the most powerful nuclear reactor ever built. The program was cancelled out and no nuclear propulsion test have been conducted since.

The best candidate for a high specific impulse nuclear propulsion system is the gas core reactor. However, some care combined with temerity would be involved with lighting this candle. Here the uranium is heated to a gas by its own nuclear reactivity. A configured gas of hydrogen is established to prevent the hot uranium from making contact with the wall of the reactor. This surrounding shell of hydrogen flows around this core and provides the rocket reaction mass. In effect this propulsion system is a sustained nuclear explosion. Quite obviously the rocket plume would drag a fair amount of the radionucleides out with them. To prevent or reduce this various shields have been proposed, such as a quartz container. However, so far there does not appear to be a material capable of such sustained containment. Further, a fair amount of detailed balance would be required to control this thruster. Yet this propulsion system could in principle deliver a specific impulse of $s \simeq 3000$ sec. No testbed system of this type has been constructed or tested. Radiation safety and environmental concerns with this have blocked any serious consideration for its development.

In between the solid core and gaseous core nuclear propulsion options are various architectures for liquid core systems. These systems have specific impulses is the 1000 to 2000 sec range. This is also expected to have some radioactivity in its rocket plume. If this were developed it, as well as the gaseous core thruster would have to be run outside the magnetopause, a region near Earth of trapped charged particles in the planetary magnetic field, so as to prevent contamination on Earth. Again for safety and environmental concerns no serious program has been initiated for development of this.

There is then a form of nuclear propulsion that might be called outrageous. The earliest concept suggested by Stanislaw Ulam had small nuclear bombs exploding at the rear of a spacecraft. Project Orion was a design study carried out by General Atomics in the late 1950's and 60's. At the rear of the craft is a pusher plate, constructed with large amounts of steel, the explosion pushed on in accordance with Newton's third law of motion. A series of nuclear detonations would then push the craft forward in a sort of "putt-putt" fashion. The pusher plate would be attached to the body of the craft by large shock absorbers. For nuclear explosives as shaped charges that direct their explosion in a dipole fashion, instead of in a spherically symmetric release of energy, this could efficiently produce a specific impulse $s = 6000$ sec. The original design was an interplanetary ship with a crew of 100 to 200 that could reach the planets in a months instead of years.

While the system appeared workable in principle it was cancelled in 1965 due to the Treaty Banning Nuclear Weapon Tests in the Atmosphere, in Outer Space, and Under Water. Further, the use of fission devices means this spaceship would leave radioactive pollution in its wake, which near Earth would result in a measurable increase in radioactivity. Exponents of this idea propose that cleaner fusion devices, hydrogen bombs, be used instead. However, detonating a fusion bomb requires a fission bomb, so the problem remains. Currently this idea is politically moribund or dead.

A related concept was advanced with Project Daedalus [3.4]. The British Interplanetary Society conducted Project Daedalus to design a concept for an interstellar un-piloted spacecraft. The craft could reach a nearby star within a human lifetime. The system is similar in concept to Orion, but the nuclear explosions were due to the fusion of deuterium-lithium pellets.

$$D +^{6}\text{Li} \rightarrow 2^{4}\text{He} + 22.4\ \text{MeV}. \tag{3.3}$$

This is a highly efficient fusion reaction to use, for there is no loss of energy in neutrons and all the energy is in charged species which can be manipulated electromagnetically. Large lasers or electron beams would implode lithium and deuterium to induce the fusion by inertial confinement. This lack of fission explosions removed the radionucleide problem with the Orion concept. The cavity for this fusion would employ magnetic fields to funnel the hot plasma helium out the back of the ship to prevent damage to the cavity walls.

This program was conducted at a time when nuclear fusion, in particular inertial confinement, appeared to have considerable promise. However, nuclear fusion power appears as remote now as it did then. The project never went beyond this theoretical level. Yet if nuclear fusion by inertial confinement is developed into a working power system this may reemerge as a reasonable propulsion system. It would likely provide an efficient propulsion

Fig. 3.4. Artist conception of the Orion spacecraft.

system for interplanetary exploration, with some prospects for interstellar probes.

The Variable Specific Impulse Magnetoplasma Rocket (VASIMR) is an extension of the ion thruster [3.5]. The thruster uses a plasma instead of a beam of ions. The neutral plasma will not suffer from beam separation issues seen with an ion beam. As such the density of the plasma beam can be much larger. This will permit higher thrusts than the ion rocket. A plasma is accelerated and controlled by electromagnetic and magnetic fields. By adjusting the density of the plasma the parameters of the thruster may be controlled. The specific impulse of the VASMIR is estimated to be in the 1,000–30,000 sec. The low end of this scale puts it in the range of the ion rocket, but for plasma density larger than an ion beam. This should provide several times the thrust of the ion rocket. Controlling the manner of heating and a magnetic choke for plasma flow, VASIMR adjusts the exhaust rate of flow and speed.

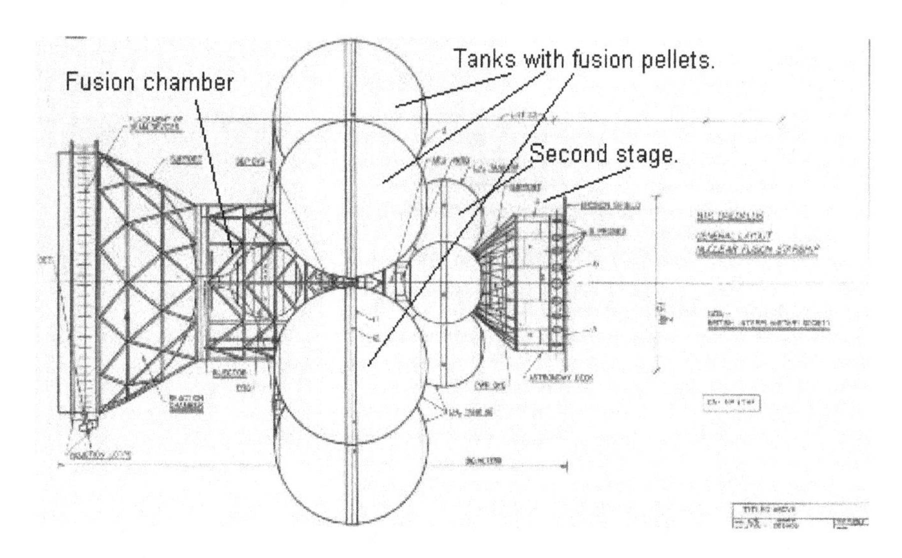

Fig. 3.5. Schematic for the Daedalus space probe.

The reaction mass or propellant is hydrogen. The application of an electromagnetic fields to H_2 molecules excites the energy of electrons, causing them to become more energetic and leave the molecule. This produces a plasma of protons and electrons. The charged particles in the presence of a magnetic field will execute circular motion around the field lines given by the Lorentz equation

$$F = q\mathbf{v} \times \mathbf{B}\,, \tag{3.4}$$

for q the electric charge, $\mathbf{v}$ the velocity of the electron and $\mathbf{B}$ the magnetic field vector. The cross product insures that the force on the electron is perpendicular to its velocity and the magnetic field. The use of the centripetal acceleration $a = m\omega^2 r$ in Newton's second law determines the cyclotron frequency, the frequency the charge cycles around the magnetic field line, is then

$$\omega = \sqrt{\frac{qv_\perp B}{mr}}\,, \tag{3.5}$$

for $v_\perp$ the component of velocity perpendicular to the magnetic field. An electromagnetic field tuned to this frequency increases the energy of the ions and heats them to 10^7 K.

Obviously this system must be supplied electricity by some power supply. This most likely would be from a nuclear reactor to meet the power

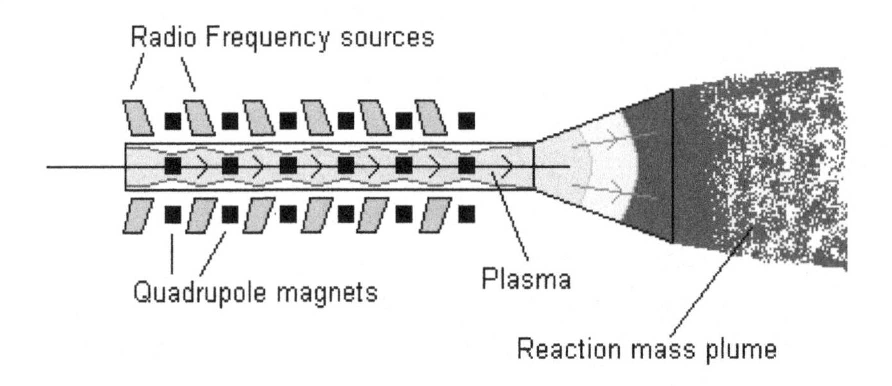

Fig. 3.6. Schematic for a VASIMR propulsion unit.

demands of the VASIMR. So far such power requirements are formidable, requiring large power supply devices. So far this system is a laboratory system and has not seen any space application.

The final mode of propulsion is the solar sail [3.6]. This has been set for last, for this is technically not a rocket. The craft does not expel any reaction mass, nor does it need to generate large amounts of energy. As such these are beneficial aspects to this, and the physics is completely different than the previous propulsion systems. A large panel of thin material is impacted by photons from the sun. This will exert a pressure on the sail and provide an exterior thrust to the craft. Recently a test craft, Cosmos-1, failed to reach orbit due to failure of the conventional launch vehicle. However, it is likely there will be a subsequent test.

A photon has an energy given by $E = h\nu$, where h is the Planck unit of action and ν is the frequency of the photon. Energy is defined by a force displaced through a distance. Force has units of $N = \text{kg-m/sec}^2$, and energy defined as the displacement of a force $E = \int \mathbf{F} \cdot d\mathbf{x}$ has the units of $J = Nm$, called a Joule. The momentum of this photon is given as $p = E/c$, which is stated now as a fact of special relativity to be investigated further on. Pressure is a force per unit area, where as seen in the discussion of Newton's law $F = dp/dt$. Thus a pressure due to photons is of the form

$$P = \frac{1}{Ac}\frac{dE}{dt}\,. \tag{3.6}$$

Here dE/dt is the power, with the unit of watts $\text{W} = \text{J/sec}$, and by dividing the power by an area A this energy is incident on this defines the energy flux density, or irradiance. The solar irradiance on the Earth is $1370\ \text{W/m}^2$.

By dividing this by the speed of light the photon pressure on a surface perpendicular to the direction of light propagation $P = 4.6 \times 10^{-6}$ p, where $p = N/m^2$ is a unit of pressure called a Pascal. This calculation assumes that the photons are absorbed by the surface instead of reflected. This means that a photon absorbing surface of one square meter will experience a tiny force of 4.6×10^{-6} N, and a surface of one square kilometer will experience of force of 4.6 N. If the photons are reflected by the surface Newton's laws indicate that the pressure on the surface is twice this amount. This is the basis of the Nichols radiometer, seen as those glass bulbs containing a vain balanced on a needle that rotates in the presence of light. Here one side of each arm of the vain is reflecting and the other side is colored to be absorbing. The pressure difference between the two sides of the thin metal piece is what causes the system to rotate. This is a windmill analogue of the solar sail.

Obviously a solar sail has to span a large area to capture as many photons as possible. Further, the material has to be made as thin as possible. A parachute designed craft that carries its payload behind it is the most obvious first idea here. For a craft sent to explore the near solar environment, a likely exploration use, a problem is that the payload is not shielded by the sail. An alternate design is to rigidify the sail with struts or masts. These may be placed opposite to the photon collecting side to protect them and the payload may be similarly shielded. The sail material is proposed to by micron thin sheets of mylar, aluminum or Kapton.

As a model situation assume solar sail made with micron thin material with a density of 1 grams per cm^3. This means that the material will have a mass of 1 g (0.001 kg) for each square meter. Thus a sail with an area of 10^4 m^2 would have a mass of 10 kg. This sail will then receive near Earth a force of 4.6×10^{-2} N. By Newton's second law $F = ma$ it is easy to find that the acceleration would be 4.6×10^{-3} $\mathrm{m/sec}^2$ or 0.46 $\mathrm{cm/sec}^2$. This puts this solar sail in the same acceleration "ballpark" as the ion propulsion system. Of course closer to the sun this can be improved due to the higher solar irradiance. If by some means the Earthly solar irradiance may be sustained on the solar sail, say by a large space based Fresnel lens or a laser, Galileo's little equation $v = at$ indicates that for 10^7 seconds the velocity increased by 10^4 m/sec or 10 km/sec, and so for three years of sustained acceleration a final velocity of $\simeq$ 100 km/sec is possible.

So far this has focused upon propulsion systems that obey Newtonian mechanics. The VASIMR thruster might be able to accelerate a craft to 5000 km/sec and the Daedalus probe might reach ten times this final

Fig. 3.7. A basic square solar sail craft depicted in the near Earth environment.

velocity. The upper range of these velocities relative to the Earth push the envelope of Newtonian mechanics, for small relativistic effects will manifest themselves. However, Newtonian mechanics works well enough in this domain. All of these options present far superior propulsion methods than chemical rockets for interplanetary exploration. Only the Daedalus concept approaches the prospect for interstellar exploration, which is the main topic of this book. Hence it is clear that propulsion technology that surpasses these will be required to send a spacecraft to another star. This spacecraft will have to approach some significant percentage of the speed of light in order to reach a star some 5–50 lightyears distant from Earth in a timely manner.

Chapter 4

Power Systems for Spaceflight

A spacecraft must be powered, otherwise it can't accelerate or is useless. Electromagnetically driven propulsion systems require a reliable power supply. The ion rocket and VASIMR are likely to be employed over the direct nuclear rocket for reasons of radiation safety. Large spacecraft sent to deep space this will require a power source that is considerably superior to what currently exists. Current power systems are solar photovoltaic generators and weak decay reactors that are not true nuclear reactors. Photovoltaic systems work well for small craft that travel close enough to the sun. The RTG radioactive type of decay reactors have been employed on craft sent to the outer gas giant planets.

It is likely that some form of nuclear energy will have to be used to send spacecraft by propulsive means to speeds of 100 km/sec or higher to the outer planets. Solar energy may work well for the inner planets. However, the energy requirements for reaching Mercury are comparatively high and a high specific impulse propulsion system must be powered to reach the planet in a timely manner. Solar energy may work well enough for this domain of space exploration. However, solar radiation at the gas giant planets is a fraction of the irradiance here on Earth. Solar photovoltaic cells would have to be inordinately large.

The use of nuclear energy in space is controversial. Many express concern over the risk of contamination if there is an accident, such as if nuclear material is released in a crash during the launch phase. A similar concern exists over putting nuclear powered satellites in Earth orbit. In low Earth orbit small amounts of atmospheric drag can bring the craft back to Earth and release radioactive materials. This happened in 1978 to a Russian nuclear powered satellite Cosmos 954 which fell in Canada. These matters doubtless can't be ignored. The launch of such materials has to have a degree of safety assurance so that they are not released in an accident. It is further probably unwise to park nuclear reactors in Earth orbit. Yet if we are to explore the outer planets it is likely that nuclear reactors will be required to power their propulsion systems.

The most common method for powering spacecraft is with solar photovoltaic cells. These have their origin with Albert Einstein who recognized that electrons generated from photons interacting with a metal were an indication of the particle-like nature of photons. This result was important in the development of quantum mechanics. The photovoltaic cell employs two materials with different materials with different potential functions for electrons. An electron excited in one material is drawn to the other through a circuit. This is the basis for the p-n junction photovoltaic cell.

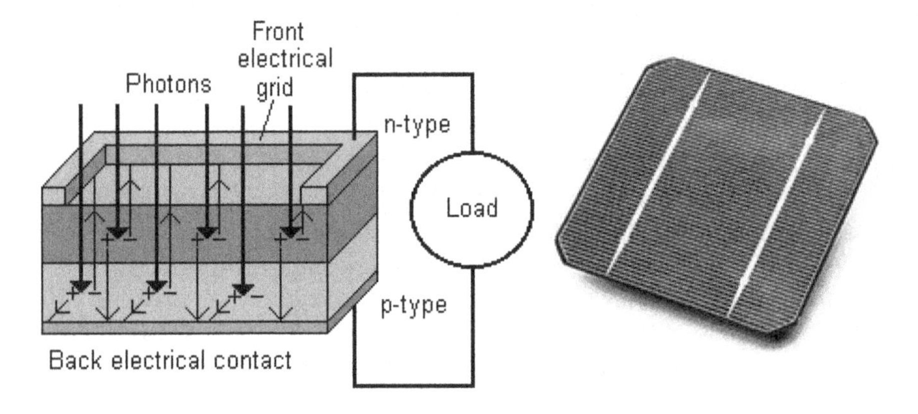

Fig. 4.1. The left is a schematic of how photons induce an electron-hole charge separation in an n-p junction and the operation of a solar cell. The right is photograph of an actual photovoltaic cell.

The most common photovoltaic cell is the silicon cell [4.1]. Silicon (Si) is a group 14 atom by virtue of its placement on the periodic table. A Si atom has 4 valence electrons in its outer shell, just as carbon. Silicon atoms may covalently bond with each other to form a crystalline solid with a long range ordering of atoms in three dimensions. Si may also form an amorphous solid with no long range order. The scale for this crystalline order defines terms for the crystalline structure of silicon; poly-crystalline, micro-crystalline, nano-crystalline depending upon the size of the crystal "grains"which make up the solid. Solar cells are constructed from each of these types of silicon, the most common being poly-crystalline. Silicon is a semiconductor, which has certain bands of energies electrons exist in, and electrons are forbidden to exist in an energy state between these bands. The two main bands are the conduction band for electrons that flow between the Si ions and the valence band of electrons bound to the Si ions. These forbidden energies are called the band gap, which requires quantum mechanics to understand fully. This will be avoided here. However, the existence of band gaps is central to the physics of semiconductors.

In a solid the electrons involved with the conduction of a current have energies in the Fermi level, which obeys strange physical properties due to quantum mechanics. However, for silicon this Fermi level is forbidden. This makes silicon a poor electrical conductor at ordinary temperatures. To improve the conductivity of silicon impurities are introduced, or "doped," with very small amounts of atoms from either group 13 or group 15 of the periodic table. These dopant atoms take the place of the silicon atoms in a few lattice slots in the crystal, and bond with their neighboring Si atoms with analogous electronic interactions the Si atoms exhibit. Group 13 atoms have only 3 valence electrons which results in an electron deficiency, and group 15 atoms have 5 valence electrons which results in an electron excess. Since the group 13 atom bonds to the crystal in the same way as silicon does this results in a positive hole. Similarly the group 15 atom gives an excess electron. The holes and electrons may then move freely around the solid and in so doing change the conduction properties of the solid. Silicon with an excess of holes is a p-type semiconductor and silicon with an excess of electrons is defined as n-type. Common group 13 atoms are aluminium and gallium, and common group 15 atoms are phosphorus and arsenic. Both n-type and p-type silicon are electrically neutral, for they have the same numbers of positive and negative charges, yet n-type silicon has negative charge carriers and p-type silicon has positive charge carriers.

A solar cell consists of a layer of an n-type silicon on a p-type silicon layer. Photons will, as Einstein explained the photoelectric effect, cause the respective charge carriers, electrons and holes, to jump into p-type and n-type silicon respectively. This is an energetically excited state. A circuit between the two silicon layers permits the electrons and holes to travel back to their respective silicon types. The excess energy is then extracted by attaching a load to the circuit. More specifically, the electrons in the valence band, those electrons that bind the Si atoms together absorb a photon and are excited into the conduction band of the p-type silicon. The presence of holes in the conduction band decrease the energy gap between the valence and conduction bands. Similarly photon absorption in the n-type silicon produces holes in its conduction band. This results in the generation of electron-hole pairs that are quantum mechanically correlated with each other. This energy imbalance is restored by permitting a current to pass between the layers. This p-n junction is the basis for the diode as well.

The efficiency of solar cells is 10–15%. Thus for a solar irradiance of 1000 W/m^2 a meter square of photovoltaic will light a 100 to 150 watt bulb. Efficiencies have improved some and the production costs have declined. It is likely that solar photovoltaics will be an increasing aspect of the electrical generating infrastructure around the world in the 21st century. The initial use of solar photovoltaic cells was with satellites, where early high costs could only be justified for space applications. Earth orbiting satellites and spacecraft sent within the inner solar system have almost exclusively employed solar panels. For spacecraft sent to the outer solar system radioisotope thermal generators must be used. Currently solar power most often generates electricity for systems on board the craft. However, solar power could also drive an ion thruster for a craft in the inner solar system. There is some interest in generating large amounts of power by solar photovoltaics for use on Earth. This energy is to be converted to microwaves and beamed back to Earth for their collection and electrical power distribution.

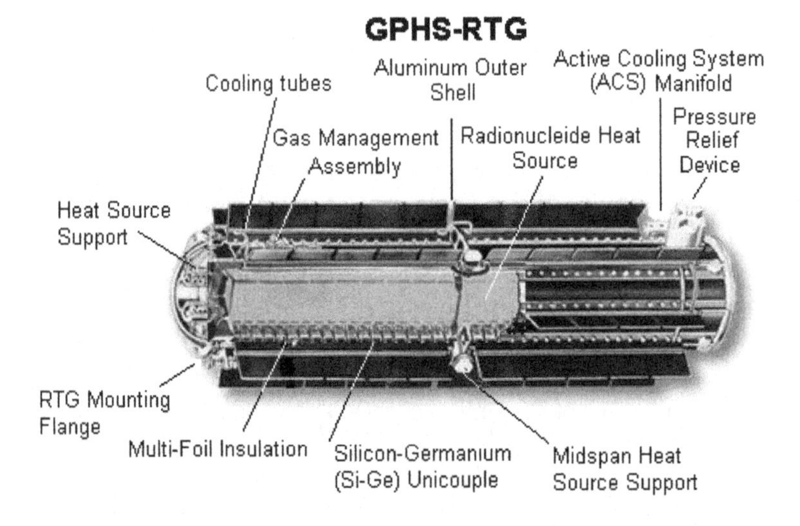

Fig. 4.2. Detailed diagram of the RTG used on the Cassini spacecraft.

Currently for deep space application radioisotope thermoelectric generators (RTG) are employed [4.2]. The concept employs a piece of radioactive metal in contact with thermocouplers which convert the heat of the material into electrical power. This is not a true nuclear reactor, but only relies upon the radioactive decay of a radionuclide. The heat difference between the radioactive material and the cold of space form the energy difference. This temperature difference is analogous to a waterfall with a paddle wheel, from which energy may be derived. The radioactive material form a core in a metal container that conducts heat. Thermocouplers on this heat conducting wall absorb this heat. The temperature difference with the other end of the thermocoupler, a heat sink, permits a heat flow through this device, where some of this heat energy may be converted to electrical energy. The thermocoupler is an infrared analogue of the photovoltaic cell. It is a p-n junction, where one junction is heated and the other attached to the heat sink.

The radioactive material must decay slowly to have a half-life long enough for the duration of the spacecraft mission. Further, the decay process must either be by a nuclear or weak interaction, which produce α (alpha) particles (helium nuclei due to nuclear decay) or β (beta) radiation (electrons or positrons due to weak decay) respectively. Some nuclei exhibit electromagnetic decay processes, due to a "reshufflling" of charges in the nucleus. This results in high energy photons called γ (gamma) rays that are highly penetrating and thus less desirable. Even still the charged alpha and beta radiation scatters with charges that make up the nuclear material to produce secondary X-rays. This process is called bremsstrahlung radiation, which occurs when a charge is accelerated. Radioactive processes often have a small neutron emission level. Neutrons can damage materials over time. As such the container around the radioactive material must be shielded to protect the thermocouplers and other components from damaging exposure. This places constraints on the types of radioactive material that can be used. The most acceptable is ${}_{94}\mathrm{Pu}^{238}$ which has the longest half-life and the lowest shielding requirements. Only 1/10 of an inch of material is need for shielding of ${}_{94}\mathrm{Pu}^{238}$, where the container is often all that is required. ${}_{94}\mathrm{Pu}^{238}$ has a half-life of 87.7 years and is a low gamma and neutron emitter. This means that the RTG will have a power loss by a proportion $1–0.5^{1/97.7}$ every year. The RTG employed on the Voyager craft were rated at 470 W, which means this has dropped to 80% since launch. This is further compounded by the degradation of the thermocouplers as well.

RTGs are only about 5–10% efficient. This means that over 90% of the radionucleide generates unused heat energy. This is a factor in launch

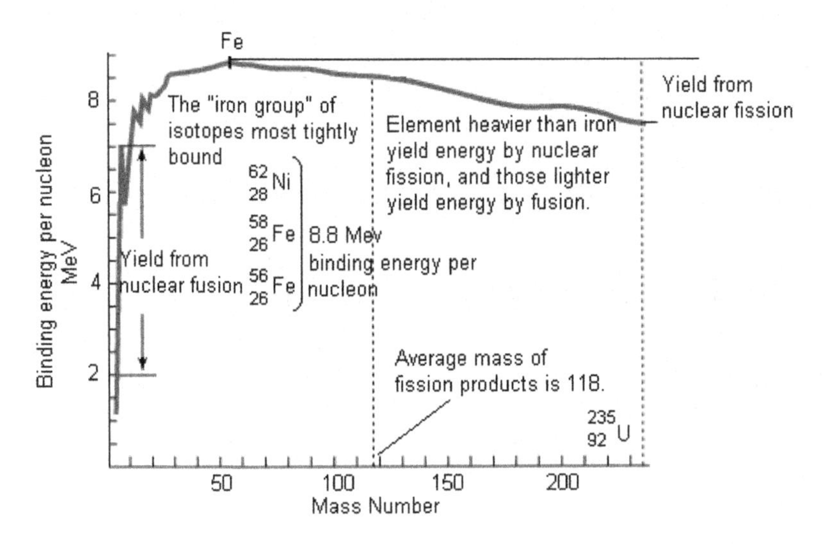

Fig. 4.3. The binding curve of energy for nuclei.

weight constraints. However, RTGs are relatively cheap to construct and plutonium is available from breeding reactors. RTGs do not have the controlled reactor issues that would arise with a sustained nuclear fission system on a spacecraft. They have then been the optimal compromise for spacecraft with moderate power requirements.

A common application of RTGs is as power sources on a spacecraft sent far into the outer solar system. Spaceprobes that travel far enough from the Sun, the inverse square law gives a $\sim\ (r_0/r)^2$ factor in irradiance decrease, where solar panels are no longer viable. The deep space probes Pioneer 10, Pioneer 11, Voyager 1, Voyager 2, Galileo, Ulysses and Cassini employed RTGs [4.3]. RTGs were used to power the two Viking spacecraft to Mars and powered scientific experiments left on the Moon by the crews of the Apollo 12, 14–17 lunar missions. Apollo 13 carried an RTG, but the lunar lander was used to carry the crew back to Earth after the service module failed. The RTG, or pieces of it, ended up in the ocean Tonga trench.

The Galileo and Cassini missions generated unpopular attention by some over the launching of nuclear materials into space. Considerable protest was made over this. The protests were in some ways high on drama and low on technical reality, but these protests can't be ignored. The concern at large is over the use of nuclear power in space, issues of launch safety and the obvious connection between space nuclear power and military applications.

The SNAP-10A was the only true nuclear reactor used in space by NASA. The Russians employed nuclear reactors with their RORSAT program. The SNAP-10A reactor powered a satellite in low Earth orbit in 1965. However, after 43 days a voltage regulator failed, which caused the reactor to be jettisoned into high Earth orbit. Nuclear power in space has become a flashpoint for those on the anti-nuclear side of the nuclear debate.

It is well known that a nuclear reactor maintains a self-sustained nuclear chain reaction. This approach was pioneered by Enrico Fermi and Leó Szilárd in 1942 at the University of Chicago. This first nuclear reactor, sometimes called a pile, was a step in the development of the atomic bomb during World War II. It demonstrated a sustained nuclear fission process with uranium was possible. Neutrons which fissioned the uranium were moderated with the use of graphite. The trick with a sustained nuclear reaction is to prevent the exponential runaway process of a nuclear chain reaction from resulting in a sudden release of energy. A nuclear bomb conversely exploits this. Fuel rods of uranium or plutonium are suspended in some moderating material, graphite or water, to slow neutrons. Control rods are inserted between the fuel rods to absorb neutrons and are inserted into the pile or withdrawn to maintain the nuclear reaction at a controlled rate. Of course for a nuclear reactor that produces large amounts of energy complex cooling systems are required, which must still operate if the reactor is shut down for there is still heating from nuclear decay processes.

The most elementary nuclear reaction is the absorption of a neutron by ${}_{92}U^{235}$ to form an transient unstable isotope ${}_{92}U^{236}$ [4.4]. This isotope

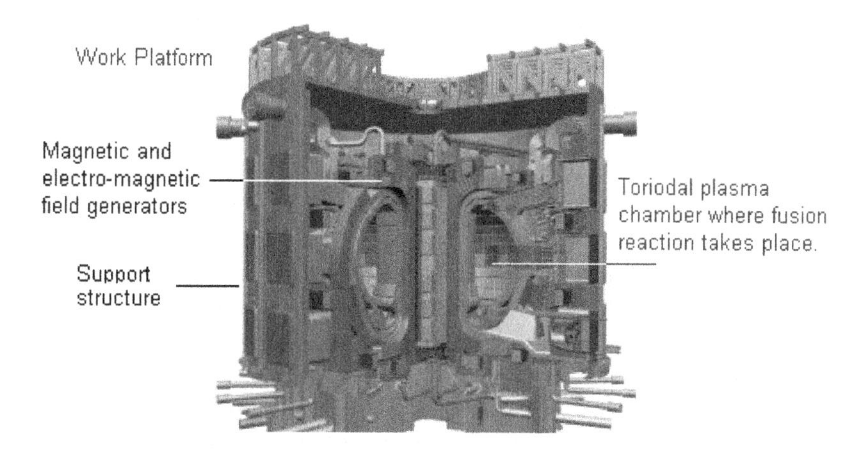

Fig. 4.4. The ITER Tokamak fusion reactor. http://www.iter.org/copyright.

fissions to produce up to three neutrons and fission daughter products. Some γ rays are produced by the rearrangement of charged species in the reaction, or by a secondary electromagnetic interaction. The fission daughter products have a high velocity and interact electromagnetically with other nuclei. The relative accelerations between these charged nuclei results in bremmstrahlung radiation in the form of γ ray and X-ray photons. Typically the fission daughter nuclei have atomic weights of 100 AMU and 132 AMU. The binding potential per nucleon, a proton or neutron is 7.5 MeV, MeV = million electron volts, for ${}_{92}\mathrm{U}^{235}$. An electron volt eV is 1.6×10^{-19} j of energy. If the nucleus is split into two nuclei of equal mass, from the average binding energy per nucleon the energy difference in binding energy between ${}_{92}\mathrm{U}^{235}$ and its fission products is (8.4–7.5) 235 MeV or 211 MeV. The energy is mostly carried by the nuclear fission products $\sim$170 MeV, about 5 MeV is carried of by neutrons and the rest is in γ rays and the radioactive energy of the daughter products. An average chemical reaction between two molecular species is around 10 eV, so the nuclear fission produces around 10 million times the energy per elementary interaction. To convert this energy to its heat equivalents, a single gram of ${}_{92}\mathrm{U}^{235}$ produces 2×10^{10} cal, while a gram of carbon, the main constituent of coal, produces 7827 cal. A nuclear reactor that consumes a ton of uranium in a year will in principle generate as much energy as a coal fired plant which consumes over a million tons of coal. The ratio between the size of a bomb and its yield for chemical and nuclear explosives scales similarly.

The ${}_{92}\mathrm{U}^{235}$ nucleus is readily fissionable by slow, or thermal neutrons. Thermal neutrons refer to neutrons slowed by reactor moderators. This was the first isotope of uranium considered for nuclear energy and the atomic bomb. However, this isotope composes only .7% of uranium isotopes. The remaining percentage is largely ${}_{92}\mathrm{U}^{238}$, which is not fissionable by slow neutrons. Yet it was found that ${}_{92}\mathrm{U}^{238}$ produced two new atomic elements not previously identified. This reaction is

$$ {}_{92}\mathrm{U}^{238} + n \rightarrow {}_{92}\mathrm{U}^{239} \rightarrow {}_{93}\mathrm{Np}^{239} + \mathrm{e}^{-} \rightarrow {}_{94}\mathrm{Pu}^{239}\,. \tag{4.1} $$

The two new atomic elements are neptunium (Np) and plutonium (Pu), which do not occur naturally on Earth. ${}_{94}\mathrm{Pu}^{239}$ fissions with slow or thermal neutrons, which made it the optimal nuclear fuel for reactors and for bomb making material. For reactors plutonium will fission in a stable manner. However, for bombs it was found that a quick exponentially exploding chain reaction could not be achieved unless the material was rapidly compressed. Hence high explosives are used to implode the material as it receives neutrons from a source trigger.

For space power applications it is obvious the energetic advantages are clear [4.5]. As a simple model a spacecraft with 1000 kg of chemical propellant is compared to a spacecraft with the same 1000 kg of reaction mass propelled by an ion thruster or VASIMR thruster powered by a nuclear reactor. For a small nuclear reactor there might be about a 10% energy conversion from nuclear energy to the kinetic energy of the reaction mass and spacecraft. This is due to thermodynamic losses of heat during energy conversion steps. If the reactor has 1 kg of fissionable material, this has as much energy as about 10^6 kg of chemical propellant. Thermodynamic losses reduce the usable portion of this by a factor of 10. Kinetic energy of a body with a mass m moving at a velocity v is $K = \frac{1}{2}mv^2$. The nuclear powered spacecraft then has about 10^5 times as much kinetic energy available. So if the chemical rocket may change the velocity of the spacecraft by Δv, often called "delta vee" the nuclear powered craft has a Δv as much as 316 times as large. Thus a nuclear powered VASIMR craft could reach velocities of $\sim$3,000 km/sec, compared to a chemically propelled craft of comparable mass and reaction mass with a $\Delta v \sim 10$ km/sec.

For interstellar travel this is still hopelessly slow. This example is about one percent the speed of light, where this might at best be improved by a factor of 10 for any fusion powered spacecraft. This is why nuclear energy simply can't send a spacecraft to a nearby star within any reasonable time frame. Yet, nuclear power provides clear advantages for interplanetary spacecraft are apparent.

Concerns over the safety of launching actinide materials from the Earth exist. It raises the question of whether nuclear fusion might be used instead. As yet controlled nuclear fusion that offers a net energy output has not been achieved. Yet it is possible that in the 21st century this may change. If nuclear fusion is obtained and employed in spacepower or propulsion the advantages could improve by a factor of 10 or more.

Nuclear fusion is the opposite of nuclear fission, in that light nuclei are forced together into heavier elements and energy is released. Fission is the splitting of heavier nuclei with the release of energy. The nuclear process exhibits an energy output for fusion for very light elements and conversely energy produced by fission requires heavy elements. Iron defines the middle point, where for elements lighter than iron more energy is required to fission them than the energy liberated, and the fusion of nuclei to form elements heavier than iron is also an energy absorbing process. Such processes are endothermic, while nuclear processes of fusion and fission at the opposite ends of the periodic table for light and heavy nuclei respectively are exothermic. This binding curve of energy is mostly an empirically known aspect of nuclear physics. Yet it is known that light nuclei under tremendous heat and pressure will fuse into heavier nuclei and release energy. This is the process which powers the sun and other stars.

The first application of nuclear fusion was the hydrogen bomb first proposed by Edward Teller. The bomb was developed by Teller and Stanslaw Ulam, where Ulam solved some of the most important problems, in the late 1940's and detonated in 1950. This development was during the ramp up of the cold war between the United States and the Soviet Union. Isotopes of hydrogen are driven together to counter their mutual electrostatic repulsion by heat and pressure to initiate the nuclear fusion reaction. The hydrogen bomb requires a small fission explosive, called a plutonium trigger, to initiate this. This leads to the idea that if a plasma of hydrogen isotopes of deuterium and tritium could be sustained at high enough temperatures and pressures that they might fuse into helium to produce energy. This program ran right away into difficulties of maintaining a plasma in a stable configuration. The standard fusion reaction is between a deuterium nucleon, an isotope of hydrogen consisting of a proton and a neutron $D = {}_1\mathrm{H}^2$, with tritium $T = {}_1\mathrm{H}^3$ according to

$$D + T \rightarrow {}_2\mathrm{He}^4(3.5\ \mathrm{MeV}) + n(14.1\ \mathrm{MeV})\,. \tag{4.2}$$

This process is used in nuclear explosives, and is the favored one by controlled fusion researchers. The energy released is large, and energy of the α particle, or ${}_2\mathrm{He}^4$ which consists of two protons and two neutrons, is relatively large. These two species exist in the lowest energy level with opposite spin states. An additional nucleon must exist in a high nuclear energy level, which for helium is the unstable ${}_2\mathrm{He}^5$ isotope. The Pauli exclusion principle dictates that only one particle of half integer spin can exist in a state. This is stated here as a "matter of fact" as this is a result of quantum mechanics. The protons and neutrons in the ${}_2\mathrm{He}^4$ may then lay in the bottom of the nuclear energy potential in a minimal quantum level with opposite spins. That ${}_2\mathrm{He}^4$ is at low energy means a large energy is released. However, it has to be pointed out that this suffers from one difficulty. Most of this energy is carried away in a neutron. For a nuclear explosive this is not a problem, for this neutron will scatter about with other nuclei and thermalize the environment. In nuclear fusion this neutron can't be manipulated readily by electromagnetic means. So much energy is lost in this process, and neutrons damage materials.

Currently the two main competing ideas are magnetic confinement and inertial confinement. In the first case a plasma is confined by magnetic means and in the second by the implosive effect of converging laser beams or particle beams. Magnetic confinement, called Tokamak reactors, confines a plasma in a toroidal chamber heated by electromagnetic means. A solenoid of arbitrary length will produce a constant magnetic field within it. However one that is wrapped around into a torus has a weaker magnetic field on its outer side compared to the inner side. This field misalignment is corrected for with poloidal magnetic fields. Currently the ITER (International Thermonuclear Experimental Reactor) is the leading experimental Tokamak proposed. Inertial confinement uses large lasers or electron beams directed onto pellets containing deuterium and tritium. Neither of these systems has produced a net energy output sustained in a workable manner.

A fusion powered craft will give the same performance as the fission powered craft without the issues of radioactive safety and launch concerns. This of course assumes that a fusion system may be made small and light enough for spacecraft power. A fusion powered ion or VASIMR drive has an expected Δv one to ten times that of the nuclear fission case.

The Daedalus project takes nuclear fusion to a different level [4.6]. The propulsion system is an inertial confinement system. A rapid series of mini nuclear bursts propels the craft forward. In this case the reaction mass is the fusion products. The proposed craft will employ the fusion reaction

$$D +_3 \mathrm{Li}^6 \rightarrow 2_2\mathrm{He}^4(22.4\ \mathrm{MeV}) \tag{4.3}$$

as the main process. This is far superior to the D-T process for the daughter products are two charged particles and their is no energy loss by neutrons. The mass of each α particles is $\simeq 3750$ MeV, which is the reaction mass. Hence the ratio of the energy output in MeV to the reaction mass in MeV is .003. This is compared to the same ratio for a chemical rocket $\simeq 10^{-10}$. Hence a "ballpark" estimate of the energy increase over that expected for a chemical rocket is $\sim 3 \times 10^6$. The Δv for this fusion craft is then up to 1250 times that of a comparable chemical rocket, and the specific impulse would be on the order of 5×10^5 sec. For such a craft that initially starts out as mostly its fuel mass it is possible to reach velocities $\simeq 10\%$ the speed of light. Such a craft could get to some of the nearest stars, such as α Centuri, within a century.

As a final note on nuclear power systems systems that exist today, whether commercial, and military systems or research designs, have a certain cultural history to them. Fermi demonstrated a nuclear chain reaction by configuring an atomic pile. Nuclear energy has rested upon this design basis ever since. The whole philosophy since 1942 has rested upon the idea of a controlled self sustained nuclear chain reaction. Similarly the idea of nuclear fusion has been that high temperatures and pressures are required. However, this need not be the only way to generate nuclear energy. Clearly a beam of neutrons could be directed at a thin target of uranium or plutonium to achieve much the same results. Indeed a fast neutron source could fission ${}_{92}U^{238}$ directly. The fission products are then magnetically separated or trapped and their energy extracted. A beam of protons could accomplish much the same, but the beam would have to be at a higher energy in order to overcome the electric repulsion between it and the nucleus. Similarly for fusion a deuteron beam on a target of ${}_{3}Li^{6}$ might achieve the same result. Further, there are none of the complexities with a self sustained chain reaction in the fission case. To turn off the reactor the beam is turned off. Similarly for the fusion case it is clear that there are none of the confinement issues with a solid piece of lithium.

This completes this chapter on power systems that are either current or within the foreseeable future. It is clear that interplanetary space exploration may be greatly expanded beyond current capabilities. However, only one power source as applied directly to propulsion is marginally capable of interstellar exploration, and the rest are simply not acceptable. It appears that radically new approaches will be required to realistically explore another star system, particularly if it is found to posses a life bearing planet. Extrasolar systems and possible astro-biology will be discussed later. Yet at this point we must explore issues of space navigation and special relativity. In order to talk about travelling to another star at some significant percentage of the speed of light we need to discuss special relativity and understand it well. Just as the above discussion on space propulsion required Newton's laws an understanding of special relativity is needed to illustrate propulsion systems required for interstellar exploration.

Chapter 5

Elements of Astrodynamics

Even if one has the propulsion and power technology to send spacecraft into space this is not enough. To send a spacecraft to the moon one does not just point the rocket at the moon like a rifle pointed at a target. A spacecraft travelling anywhere in the solar system is influenced the gravity field of nearby planets and the sun. Thus the trajectory of a spacecraft is not going to be a straight line. This is most clearly the case for a satellite orbiting the Earth. So for these Newtonian velocity spacecraft some discussion of spacecraft navigation or astrodynamics is relevant.

The simplest example of this is a circular orbit of a satellite around a planet or star. Newton's second law indicates that

$$\frac{v^2}{r} = \frac{GM}{r^2}, \tag{5.1}$$

which predicts a velocity with a magnitude $v = \sqrt{GM/r}$ tangent to a circle of radius r. A similar easy calculation can give the escape velocity. The total energy of a particle is given by the kinetic energy of that particle $K = 1/2mv^2$ plus the potential energy in the gravity field. That potential energy is $V = -GMm/r$, where the force induced by this potential is $F = -\partial V/\partial r$. The total energy is then

$$E = \frac{1}{2}mv^2 - \frac{GMm}{r}. \tag{5.2}$$

If this total energy vanishes it is easy to show that $v = \sqrt{2GM/r}$. This velocity vanishes when $r \to \infty$, and so this velocity is the break even point for escape velocity. It is apparent that a spacecraft in a circular orbit at radius r with a velocity v must be accelerated to a velocity $v_{esc} = \sqrt{2}v$. A spacecraft in Earth low earth orbit has a velocity $v \simeq 7$ km/sec and the escape velocity is $v \simeq 11$ km/sec. A spacecraft given a higher velocity will escape as well, and if the sun were not present it would reach "infinity" at a nonzero velocity. The velocity of the Earth around the sun is 29.5 km/sec and the velocity required to escape the solar system is $\sqrt{2} \times 29.5$ km/s = 41.7 km/sec.

To consider the motion of a body under a central $1/r^2$ force, such as gravity, it requires that the problem be cast in polar coordinates. The center of the coordinate plane is the center of the gravitating body, and by extension is the center of force. The radial and and transverse, or angular, components of the acceleration are

$$a_r = \frac{d^2r}{dt^2} - r\left(\frac{d\theta}{dt}\right)^2, \; a_\theta = \frac{1}{r}\frac{d}{dt}\left(r^2\frac{d\theta}{dt}\right)^2. \tag{5.3}$$

For a central force the angular acceleration is zero. For some physics problems this will not always be the case, such as with the motion of a charged particle in converging or diverging lines of magnetic force. The vanishing of the angular acceleration means that

$$r^2\frac{d\theta}{dt} = j, \tag{5.4}$$

for j a constant angular momentum per unit mass. The acceleration in the radial direction is then identified by Newton's second law with the gravitational acceleration

$$a_r = -\frac{GM}{r^2}. \tag{5.5}$$

It is conventional to convert this problem to the variable $u = 1/r$. Further, since $d\theta/dt = ju^2$ the problem may be converted from a time variable to the radial component of motion with

$$\frac{d^2u}{d\theta^2} + u = \frac{GM}{j^2}. \tag{5.6}$$

Both sides of this differential equation are constant and is for an oscillating spring with a constant force. This differential equation then defines the motion of the particle by

$$r = \frac{1}{u} = \frac{j^2}{GM}\frac{1}{1 + \epsilon cos(\theta - \theta_0)}, \tag{5.7}$$

for θ_0 and ϵ constants of integration, and ϵ defines the eccentricity of the orbit. The eccentricity is given by the angular momentum by $j^2/GM = a_r(1 - \epsilon^2)$, and an elliptical orbit is determined by $\epsilon < 1$.

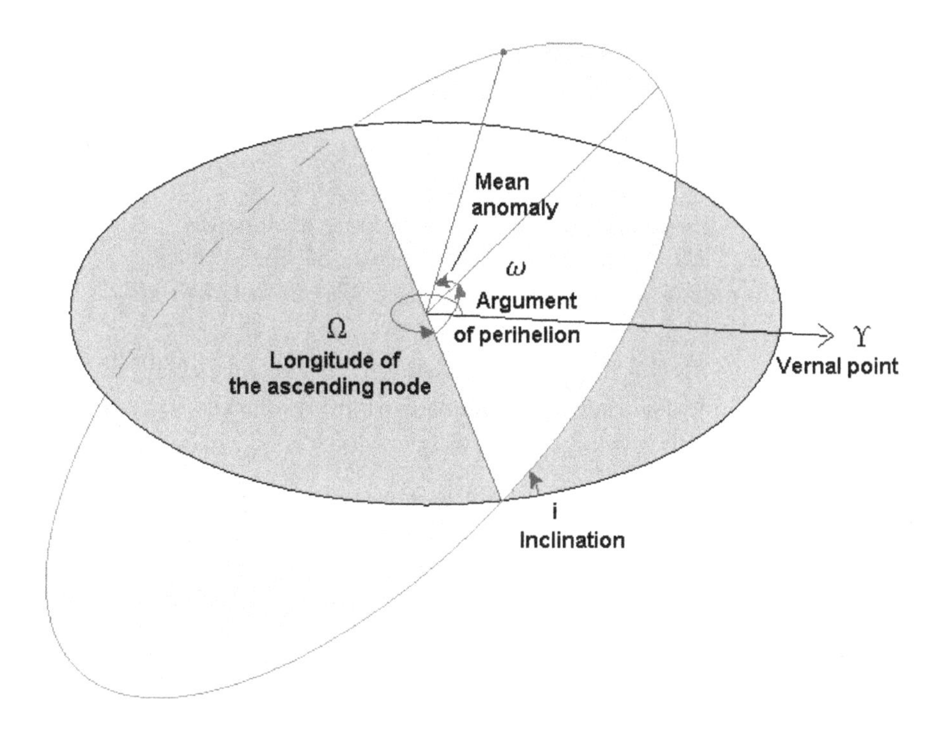

Fig. 5.1. Orbital parameters used to characterize an orbit.

A real orbit around a star or planet is elliptical, as laid down by Kepler nearly 400 years ago. With the launching of Sputnik by the USSR in 1957 this was extended to astrodynamics. Orbits in astrodynamics are defined in the same way as with celestial mechanics as ellipses described by orbital elements. The Keplerian set of orbital elements [5.1] for a elliptical orbit are:

— Epoch (time): A moment in time for which celestial coordinates or orbital elements are specified. This is usually expressed in Julian seconds.
— Inclination (i): The angle subtended by the plane of the orbit of a planet or satellite and the ecliptic, which is the orbit of Earth.
— Longitude of the ascending node (I): The angle formed at the Sun from the First Point of Aries to the body's ascending node, as measured in the reference plane of the ecliptic. The ascending node is the point where the orbiting body passes through the plane of ecliptic.
— Argument of periapsis (ω): The angle between the ascending node and the periapsis, or point of closest approach of an orbiting body to the central attracting body. For equatorial orbits this is not defined.

— Eccentricity (e): For elliptic orbits it is calculated from distance at periapsis (perigee or closest approach) and apoapsis (apogee or further approach) as

$$e = \frac{d_a - d_p}{d_a + d_p}, \tag{5.8}$$

where d_p is distance at periapsis, and d_a is distance at apoapsis.

— Mean anomaly (M) at epoch: M is a measure of time, which is a multiple of 2π radians at and only at periapsis. The mean Longitude L is defined as

$$L = M + \omega\,. \tag{5.9}$$

— Orbital period (T): The time for a complete orbital revolution as given by Kepler's third law.

The first of these is not strictly an orbital parameter, but is how one sets the reference clock. This does become a dynamical variable if velocities are large enough to exhibit relativistic effects. However, this will be ignored.

These parameters are used to define orbits for planets and satellites in polar coordinates on a plane tilted relative to a the plane of the ecliptic. The ecliptic and periapsis define this tilting. The ascending node indicates a point of intersection of the orbit with the ecliptic, and its longitude is that angle between a line from the central body to that point and a coordinate direction of the ecliptic. The coordinate direction is the first of Aries, which is the orbital position for the spring equinox on Earth. The remaining orbital parameters give the closest approach of the planet, the structure of the elliptical orbit by its eccentricity and where at a given time the orbiting body is. These six pieces of datum are equivalent to the specification of x, y, z and v_z, v_y, v_z in cartesian coordinates for the position and velocity of the orbiting body. However, the orbital parameters are expressed more naturally according to the geometry of elliptical orbits. This information is then used to integrate the orbit of an orbiting body according to Newton's second law. It is left to the interested reader to study this further.

For interplanetary travel a spacecraft is transferred from Earth orbit to the orbit of another planet. The transfer that performs this with the least energy is a Hohmann transfer orbit. This further assumes that the orbits of the two planets are on the same plane. This orbit is an elliptical orbit that "kisses" the orbit of the two planetary orbits. For a planet in an elliptical orbit the total energy per unit mass for this orbit is

$$E/m = \epsilon = \frac{v^2}{r} - \frac{GM}{r} = \frac{(GM)^2}{J^2}\Big(1 - e^2\Big), \tag{5.10}$$

where e is the eccentricity of the orbit and J is the angular momentum of the orbit. This may be further written as $\epsilon = -GM/2a$ where $a = \frac{GM}{J^2}(e^2 - 1)$ is the semi-major axis. This is an important result in classical mechanics: The energy of an elliptical orbit depends upon the semi-major axis. Hence for a planet its energy/mass is $\epsilon = -v^2/r - GM/r$ and the same energy per mass obtains for a spacecraft in this kissing orbit at its semi-major exist $\epsilon = -GM/2a_2$. It it then apparent that the delta-vee required for the orbital transfer is

$$v^2 = GM\Big(\frac{2}{r} - \frac{1}{a}\Big). \tag{5.11}$$

It is worth noting in equation 5.5 that for $e = 1$ the energy is zero, which geometrically corresponds to a parabolic orbit for the escape trajectory. For $e = 0$ the orbit is circular and the energy is maximal. For comets the eccentricity is very close to unity. It has been suggested that interstellar space is populated with small ice bodies which occasionally swing by a star to become a transient comet with $e \geq 1$.

The Hohmann transfer orbit is the minimal energy configuration [5.1]. It is largely employed for interplanetary space missions. It is also used in the reentry of manned spacecraft. A delta-vee on the shuttle will put it on an elliptical orbit that kisses its more circular orbit around the Earth. This brings it within the Earth's atmosphere which is what really breaks the velocity of the spacecraft. This also illustrates why it takes energy to send spacecraft to the inner solar system. The extreme case is a suggestion to send garbage to the sun. To do this a delta-vee of $29.5km/sec$ is required to put the payload in a dead stop relative to the heliocentric coordinates. From here the payload would then fall into the sun. This is a large change in velocity and at the extreme end of current launch vehicle capability.

For a transfer orbit to a planet on a different orbital plane than the Earth this requires a non-Hohmann orbital transfer. This requires an acceleration that changes the energy/mass by $\mathbf{v} \cdot \mathbf{a}/a^2$. This is an added complexity that will not be explored here. Obviously for more muscular thrusters the transfer is also non-Hohmann. The details of these aspects of astrodynamics are left to the reader to explore further.

Astrodynamics for star travel is comparatively simple. All one needs is to compute the average velocity of the spacecraft and the relative velocity of the target star. It is then a simple matter to figure the direction the craft should be aimed to intersect the star. Interstellar space flight involves high velocities in a region where gravity is not a significant factor.

The classical mechanics developed to this point permits us to mention why systems with three or more bodies can't be solved in closed form. For any N body problem there are 3 equations for the center of mass, 3 for the momentum, 3 for the angular momentum and one for the energy. These are 10 constraints on the problem. An N-body problem has $6N$ degrees of freedom. For $N = 2$ this means the solution is given by a first integral with degree 2. For a three body problem this first integral has degree 8. This runs into the problem that Galois illustrated which is that any root system with degree 5 or greater can't be solved algebraically. First integrals for differential equations are functions which remain constant along a solution to that differential equation. So for 8 solutions there is some eight order polynomial $p_8(x) = \prod_{n=1}^{8}(x - \lambda_n)$, with 8 distinct roots$\lambda_n$ that are constant along the 8 solutions. Since $p_8(x) = p_5(x)p_3(x)$, a branch of algebra called Galois theory tells us that fifth order polynomials have no general algebraic system for finding its roots, or a set of solutions that are algebraic. This means that any system of degree higher than four are not in general algebraic. At the root of the N-body problem Galois theory tells us there is no algebraic solution for $N \geq 3$. Galois theory is a subject outside the scope of this book, but it is the basis of abstract algebra, groups and algebras used in physics.

This reasoning is the basis for a series of theorems by H. Bruns in 1887, which lead to Poincaré's proof that for $n > 2$ the n-body problem is not integrable. For this reason the stability of the solar system can't be demonstrated in a closed form solution, but only demonstrated for some finite time by perturbation methods. In chapter 14 the results of numerical integrations of a reduced solar system with the sun, Earth and Jupiter are illustrated. The random nature of this dynamical system is illustrated, and reduced to a simple model of a random chain, or Ising model. This is then used to estimate whether known extra-solar systems might support a planet in an orbit sufficiently stable to permit the evolution of life.

Chapter 6

Special Relativity

In spite of the success of Newton's three laws of motion for two centuries some problems began to arise. These problems were the absence of an aether electromagnetic waves were though to travel on, a precession in the orbit of Mercury not accounted by perturbations from the other planets, and the inability of classical physics to account for the distribution of light frequencies by a non-reflecting hot body, called a black body. The last of these loose threads in the physical world view was pulled by Max Planck in 1895. The resolution of this black body problem was solved by assuming that radiation occurred in discrete energy packets. This lead to the concept of the quanta, or that energy states on a small scale occurred not in a continuum, but rather in discrete energy levels. Of course this lead further to the theory of quantum mechanics. The first of these threads lead to the prospect that electromagnetic radiation did not exist on a fixed reference frame. This stimulated ideas by Lorentz and Einstein to develop the special theory of relativity. The last problem was solved by Einstein's general theory of relativity. By 1915 the world view of physicists changed dramatically. While the story of quantum mechanics is fascinating in its own right, we will concentrate on the development of special relativity.

James Maxwell developed the equations which unified the electric field and the magnetic field into a single set of equations that described the electromagnetic field [6.1] in 1865. It was known previously, mostly by the work of Faraday, that a current could induce a magnetic field and that a changing electric field could produce a changing magnetic field, and the converse as well. Maxwell demonstrated how a changing current, known to be associated with a changing magnetic field, could produce a changing electric field. This changing electric field was a "displacement current" added to the Faraday equation. This unification of the electric and magnetic field resulted in the understanding of how a bundle of oscillating electric and magnetic field could propagate though space as a wave. It was further demonstrated that light is a form of electromagnetic wave. This electromagnetic field as a wave propagates through space at a speed $c \simeq 300,000$ km/sec. Further, the wave equations demand that this should be the case universally.

In spite of this physicists started asking the wrong questions. They assumed that electromagnetic waves were similar to water waves, and so there had to be a medium these wave propagated within. This gave a fixed frame to the universe for electromagnetic wave propagation. Yet it should have been apparent to any inquiring physicist in the late 19th century that there was a problem here. Newton's laws give a hint of this. The first law tells us that a body has inertia, and so the proper reference frame for observing physics is inertial, or not accelerating. The second law tells us that a body accelerates in direct proportion to the force applied, where the proportionality factor is that body's mass. Again the second law is only properly applied or observed from an inertial reference frame. The third law says that the force one body exerts on another is equal in magnitude to a force exerted on the first by the second in the opposite direction. This law of motion indicates that no matter the direction with which this is arranged the result is the same. Also by the first law this obtains no matter how fast these two bodies are travelling with respect to an inertial reference frame of observation. Further, this happens everywhere in space. The third law tells us that space is the same as measured by any inertial observer travelling at any velocity, that space is isotropic, for it does not depend upon any direction, and finally that space is homogenous so that the dynamical principles are independent of where in space one observes motion. This is referred to as the Galilean relativity principle. The imposition of some aether amounts to a violation of this relativity principle. The imposition of this aether amounts to the imposition of a preferred inertial reference frame with zero velocity. This is in fact an implicit violation of Newton's first and third laws of motion, at least with respect to the motion of an electromagnetic wave.

This apparent contradiction between Newtonian mechanics and the aether theory failed to grab the attention of physicists. However, this preferred frame would have measurable consequences. If an observer were travelling at 10% the speed of light with respect to the aether frame light propagating in the same direction would be seen to be propagating at 90% the speed of light. Similarly light travelling in the opposite direction would be seen to propagate at 110% the speed of light. Just as a swimmer crossing

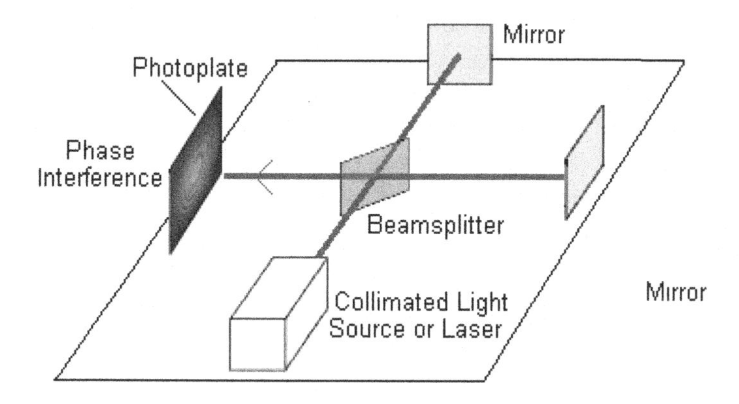

Fig. 6.1. Schematic of the Michelson-Morley interferometer.

a river is carried down stream by the current, light travelling in a direction perpendicular to this reference frame would be diverted. The consequences of this would be measurable. Since the Earth executes a near circular motion around the sun these effects would be seen at different times of the year as the Earth travels through the aether in different directions and different velocities. Of course the effect might be subtle, for the 29.5 km/sec orbital speed of the Earth around the sun is $\simeq 10^{-4}$ that of light speed. Yet by the end of the 19th century optical technology was up to the job.

Albert Michelson and Edward Morley designed just the device to do this in 1887. They floated a table in a vat containing liquid mercury, where upon the table were a light source, a half mirror beam splitter, two full mirrors and a scope. The half mirror split the light so that it would travel in two directions perpendicular to each other. Only a very narrow range of light frequencies went through this interferometer. This light would then reflect off the two mirrors and recombine at the scope. If the two light beams travelled different distances the recombined beam exhibits interference fringes where the two waves combine and cancel. Further, if as the Earth orbits the sun the speed of this light would be seen to change as the Earth travels at different directions and velocity with respect to the aether. This would mean the interference fringes would be seen to change over a year period. Michelson and Morley found no such change. The effect of the aether was absent. This cast doubt upon whether the aether existed or whether there were competing effects that masked the appearance of the aether. Michelson won the Nobel prize for this result in 1907.

George FitzGerald and Hendrik Lorentz, in 1904 and 1905 respectively, accounted for this loss of interference change by calculating how much the apparatus would have to shrink in the direction along its motion with respect to the aether. This is the famous Lorentz-FitzGerald contraction formula $L' = L\sqrt{1 - (v/c)^2}$ [6.2]. This lead to Lorentz publishing his Lorentz transformations which removed the effect of the aether. These transformations meant that lengths along the direction of motion are reduced and a clock on that moving frame is seen to slow down with respect to an observer fixed to the aether. This result is based on largely phenomenological arguments, or arguments concerning a measurement outcome within some established physical theory. The Lorentz transformations are an end result of Einstein's special theory of relativity, yet Einstein's approach was to restructure the meaning of space and time by framing it on the invariance of the speed of light.

Albert Einstein at the age of 15 read a popularization of electricity and magnetism. He learned the essential nature of electromagnetism: An electromagnetic wave consists of oscillating electric and magnetic fields that propagates at a predicted speed $c \simeq 3 \times 10^8$ m/sec. He then imagined what would happen if somebody were travelling along side the wave. This observer would see the electric and magnetic field oscillate, but remain at rest with respect to that observer's frame. However, Einstein realized that this observation would violate what Maxwell told us: Electromagnetic waves travel at the speed of light. Einstein saw a contradiction in the standard concepts of physics as a teenager, at a time when the physics community was beginning to struggle with increasingly irreconcilable problems.

Einstein realized the crux of the problem in 1905. His approach to the problem was to look at the motion of charged particles in spacetime. In particular a charged particle with an electric field will have a magnetic field circulating around it as measured by an observer in an inertial reference frame moving with some velocity relative to that particle. Einstein simply added the requirement that the speed of light must remain the same relative to all reference frames. His analysis found the Lorentz transformations, but with the concept that space and time are inter-changable. Just as a rotation of an $x - y$ axis about its origin to some new $x' - y'$ axis inter-changes the meaning of the two coordinates, a spatial direction and time can be similarly inter-changed. However, this "rotation" is a pseudo-rotation that is hyperbolic instead of elliptical (circular).

To illustrate the nature of special relativity we will consider the transformation of vectors in space plus time, or spacetime. A vector is an array of components for the values of a vector at some point. So for some arbitrarily given origin with $ct = 0,\ x = 0,\ y = 0,\ z = 0$ a vector from this origin to some other point $ct,\ x,\ y, z$ is written as

$$\mathbf{V} = \begin{pmatrix} ct \\ x \\ y \\ z \end{pmatrix}. \tag{6.1}$$

ct is used for the time coordinate as it has length units. Now assume that there is some transformation. In other words we wish to compute that vector for some other coordinate system. This gives a new vector $\mathbf{V}' = \mathbf{MV}$, where $\mathbf{M}$ is a matrix with 16 entries in a 4×4 regular or block form. The reader is advised to look at basic texts on matrix operations. The ith component of the transformed vector is

$$V_i' = \sum_{j=1}^{4} M_{ij} V_j. \tag{6.2}$$

A vector has the same length no matter what coordinate it is expressed in. To measure the length of a vector we consider $\mathbf{V}^T = (-ct,\ x,\ y,\ z)$, where the superscript T means transpose, which exchanges columns for rows. Under transposition $ct \to -ct$, because otherwise the rotation found is incorrect. The length of this vector is then

$$\mathbf{V}^T\mathbf{V} = -(ct)^2 + x^2 + y^2 + z^2, \tag{6.3}$$

which is a modified form of the Pythagorean theorem. Since the length of the vector is unchanged by any coordinate transformation this means that $\mathbf{V}^T\mathbf{V} = \mathbf{V}^{T\prime}\mathbf{V}'$ It is then apparent that

$$\mathbf{V}^{T\prime}\mathbf{V}' = \mathbf{V}^T\mathbf{M}^T\mathbf{MV}, \tag{6.4}$$

which gives a property of the rotation or transformation matrix $\mathbf{M}^T\mathbf{M} = 1$. The transpose of a matrix interchanges the rows and columns by $M_{ij} \to M_{ji}$. It is then possible to use this information to find the entries of the matrix $\mathbf{M}$.

The unchanged length of a vector under this transformation, called its invariance, means that the components of $\mathbf{V}$ and $\mathbf{V}'$ are $ct,\ x,\ ,y,\ z$ and $ct',\ x',\ y',\ z'$ respectively with

$$-(ct')^2 + (x')^2 + (y')^2 + (z')^2 = -(ct)^2 + x^2 + y^2 + z^2. \tag{6.5}$$

This problem is simplified by considering the transformation as a velocity in the x direction. This means that $y' = y$ and $z' = z$. As a trial solution consider $x' = \gamma(x - vt)$ and $x = \gamma'(x' + vt')$. The factors γ, γ' are to be determined. By reversing the roles of x and x' it is evident that $\gamma = \gamma'$. For a particle or photon moving with $x = ct$ then $x' = \gamma(c - v)t$. Further, since the speed of light is regarded as an invariant $x = ct$ implies $x' = ct'$. This leaves the two equations,

$$ct' = \gamma t(c - v), \; ct = \gamma t'(c + v) . \tag{6.6}$$

At this point it is easy to see that by multiplying these two together and dividing out the tt' term, we obtain

$$\gamma = \frac{1}{\sqrt{1 - (v/c)^2}} , \tag{6.7}$$

which is the famous Lorentz-FitzGerald contraction factor. This was derived by Lorentz and FitzGerald to contract the length of a body moving in the aether to cancel out the influence of the aether. This leaves the following Lorentz transformations

$$t' = \gamma(t - vx/c), \; x' = \gamma(x - vt), \; y' = y, \; z' = z . \tag{6.8}$$

The transformation matrix then has the form

$$\mathbf{M} = \begin{pmatrix} \gamma & -\gamma v/c & 0 & 0 \\ -\gamma v/c & \gamma & 0 & 0 \\ 0 & 0 & 1 & 0 \\ 0 & 0 & 0 & 1 \end{pmatrix} , \tag{6.9}$$

which defines the Lorentz transformation.

The γ factor is what is responsible for the time dilation and length contraction effect. An observer on one frame will observe a clock on a frame moving with velocity v tick away with time intervals slowed or increased by $\Delta t' = \gamma \Delta t$. Similarly a meter stick pointing in the direction of motion will appear shrunk by a factor $L' = L/\gamma$. There is a caveat that must be made about this. To observe another reference frame photons must "bounce off" items on that frame and reach the observer who's frame is designated as at rest. This adds complicating factors, which implies that a moving cube will appear rotated. It is also commonly said that the mass of a body is changed by γm. It is my sense that this statement is best not used, at least extensively. Vector quantities, such as momentum, transform, but scalar quantities like mass do not.

So after this bit of analysis it is good to step back and look at the big picture. This forces a modification of Newton's laws. This modification will be examined in more detail later. However, it modifies the first law of motion by saying that a body in a state of motion will remain in that state of motion if not acted on by a force as measured by an observer in an inertial reference frame, where the speed of light is invariant on all such reference frames. So not only is the motion of some massive body a constant if free of any acting force, but now on all possible reference frames the speed of light is the same. As a technical matter, the velocity of a particle is now considered in spacetime. It includes the standard spatial velocity $\mathbf{v}$, but in addition there is a time-like velocity. The spacetime velocity is written as U^a so that $U^1 = U^t$ and the components for $a = 2$, 3, 4 are the components of the spatial velocity. This modifies Newton's first law of motion by including an invariance principle. The second law must be extended to spacetime, or four dimensions, where momentum, force and acceleration are extended to four-vectors. There is further a subtle issue with what time, "t" is used in the dP/dt term, where this equation must make sense to all observers. Hence transformations between frames changes the meaning of time. The second law of motion must be reformulated accordingly. However, given these modification the relativistic second law of motion is essentially the same. The third law is similarly modified. Newton's third law indicates that space has a symmetry structure, where the physics is the same no matter where one's frame is or how it is oriented. Special relativity demands that this symmetry principle be extended to one that includes the Lorentz transformations. The symmetry of spacetime has an additional set of "rotations" that can inter-change spatial coordinates with time. These modifications are required in order to understand the physics of a relativistic rocket and photon sail.

It is clear that an understanding of dynamical principles in the framework of special relativity must be reformulated. To start this process consider the length invariance of the vector $\mathbf{V}$ above. In three dimensions the length of a vector form the origin is $|\mathbf{V}|^2 = x^2 + y^2 + z^2$. In spacetime this vector includes a ct component in four dimensions. The length of this vector is then

$$|\mathbf{V}|^2 = s^2 = -(ct)^2 + x^2 + y^2 + z^2 \,. \tag{6.10}$$

This length is often labelled by s, or τ, and is called the invariant interval, and defines the proper time. The time variable t is then the coordinate time. The negative sign on the $(ct)^2$ term is why flat spacetime of special relativity is often called pseudo-Euclidean. There are a number properties for this interval. To consider a particle travelling in the x direction let $x = vt$ and set $y = z = 0$. It is then clear that

$$s^2 = (v^2 - c^2)t^2 < 0\,, \tag{6.11}$$

which is called a time-like interval. Similarly for $x = ct$ we have

$$s^2 = (c^2 - c^2)t^2 = 0\,, \tag{6.12}$$

and is designated a null interval. It is then apparent for a particle travelling faster than light, called a tachyon, that $s^2 > 0$, which defines a space-like interval. The tachyon is a sort of fiction, or a particle that vanishes in superstring theories. It is unwise to start seeing the tachyon as some way of travelling faster than light. The invariant interval defines the proper time of the particle which travels from this origin to the tip of this vector. It is also the length that defines the path of a particle, called a worldline, in spacetime. For a particle or photon travelling the speed of light this interval is zero. A photon has no internal clock.

This structure defines a light cone. For a point defined as the origin there is a set of null lines that pass through this point. This gives the past and future light cones through this point. This defines points in the past of the origin that may causally interact with this origin, as well as points

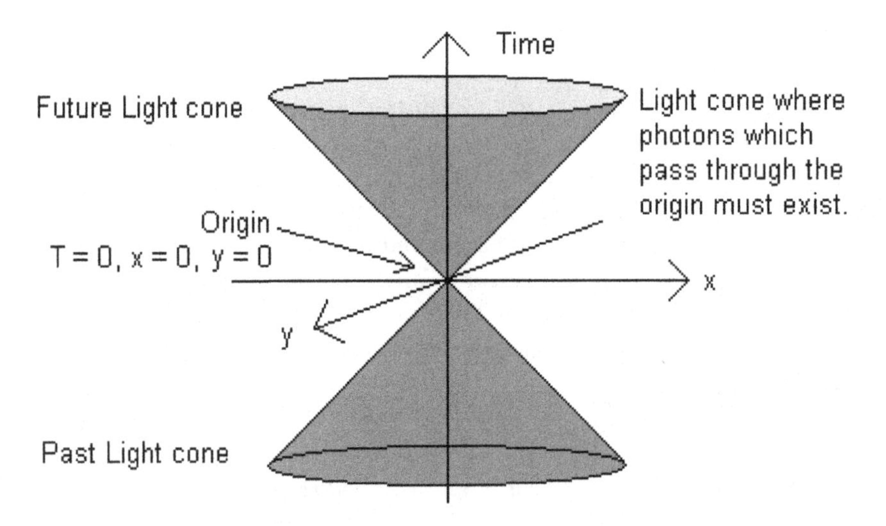

Fig. 6.2. The lightcone structure of spacetime.

in the future of the origin that may be causally influenced from the origin. The past and future events or points may interact with the point of origin must propagate on either time-like curves from points within the cone or are propagated along the cone as null lines. This is shown in Figure 6.2, where one spatial dimension has been suppressed.

A property of special relativity is that there is no universal frame from which time is measured. For instance, two points separated by a distance or ruler may have their clocks synchronized according to a reference frame at rest with respect to them. If there are flash bulbs at these points that pulse simultaneously in a frame at rest with respect to these bulbs the light pulses will meet at the center. However, for a frame moving with some large velocity with respect to this spatial distance since the speed of light is an invariant the wave fronts no longer meet at the center. This means that to this observer one bulb flashes before the other. Figure 6.3 illustrates this effect. This means that clocks can only be synchronized according to a chosen inertial reference frame.

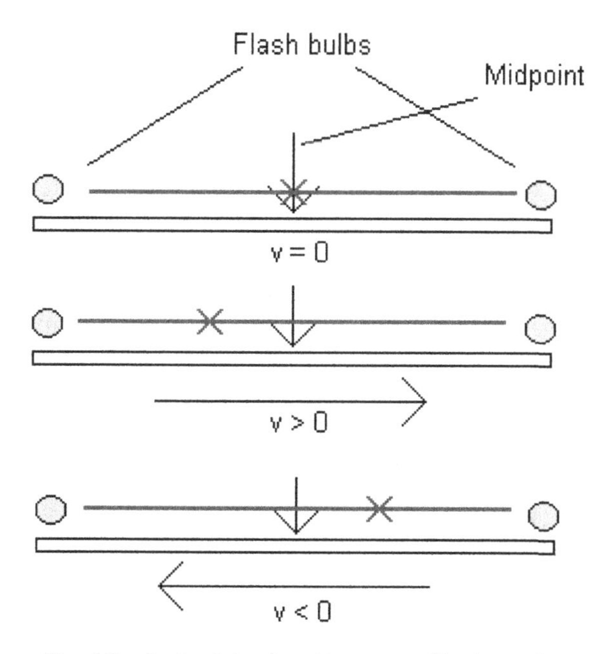

Fig. 6.3. Lack of simultaneity on spacelike intervals.

As a formalistic sideline, some books on relativity put the negative sign in the interval on the spatial terms so that

$$s^2 = (ct)^2 - x^2 - y^2 - z^2 \,, \tag{6.13}$$

which reverses the sign convention for time-like and space-like intervals. This is a matter of convention, where some people prefer to have positive intervals for time-like intervals and others prefer that the spatial part of the interval remains positive to fit with our usual three dimensional sense of things. It does not matter which convention is used, but one must stick with one through the calculation of a problem.

A similar interval exists for momentum and energy. A particle with a momentum $\mathbf{p}$ and an energy E has a spacetime momentum of

$$\mathbf{P} = \begin{pmatrix} E \\ \mathbf{p} \end{pmatrix} , \tag{6.14}$$

with $\mathbf{P}^T$ defined accordingly. The momentum-energy interval is then

$$\mathbf{P}^T\mathbf{P} = -E^2 + \mathbf{p}^2 \,. \tag{6.15}$$

What is defined on the left hand side of equation 6.14? To answer this, we first compute the Lorentz transformation of the momentum vector in spacetime $\mathbf{P}' = \mathbf{MP}$. It is left as an exercise to the reader to perform this, where the energy and momentum terms in the four vector are found to transform as

$$E' = \gamma E,\ p' = \gamma m v \,. \tag{6.16}$$

Since $\gamma \to \infty$ as $v \to c$ it is clear that light speed is an infinite barrier that imposes a fundamental speed limit in the universe. For the momentum term use the fact that kinetic energy of a particle is defined in Newtonian mechanics as the displacement of a force through a distance

$$K = \int_0^x \mathbf{F} \cdot d\mathbf{x} \,. \tag{6.17}$$

For simplicity, we consider this motion only in one direction so that the vector notation may be suppressed. We further use the fact that $F = dp/dt$ so that

$$K = \int_0^t \frac{dp}{dt} v dt = \int_0^p v dp \,. \tag{6.18}$$

Now substitute $p = \gamma mv$ and perform the integration, which is left as an exercise to the reader. The result is then

$$K = mc^2(\gamma - 1)\,. \tag{6.19}$$

This kinetic energy equation contains the famous $E = mc^2$ result. This is used in the invariant interval for the momentum-energy vector $\mathbf{P} = (E,\ \mathbf{p})$ for $\mathbf{p} = 0$, $\mathbf{P}^T\mathbf{P} = (mc^2)^2$. This interval remains invariant under all Lorentz transformations, and so

$$(mc^2)^2 = E^2 - |\mathbf{p}|^2\,. \tag{6.20}$$

At this stage we have the Lorentz transformations of special relativity, the nature of the invariant interval and now the invariant momentum-energy interval.

To extend this theory, the interval needs to be cast in a differential form. A straight line in spacetime defines a particle without the action of any force upon it, which conforms to the spacetime extension of Newton's first law. However, in order to generalize Newton's second and third laws it is necessary to consider ds expressed according to dt, dx, dy, dz as

$$ds^2 = -(cdt)^2 + dx^2 + dy^2 + dz^2\,. \tag{6.21}$$

For a particle on a spacetime path in the presence of a force, $s = \int ds$ gives the invariant proper time on the particle world line. Since ds is an infinitesimal element of the invariant interval, and length of a path in spacetime, this is the appropriate time to be used in a Newton's second law calculation. The four dimensional force vector on a particle that is accelerating is

$$\mathbf{F} = \frac{d\mathbf{P}}{ds}\,. \tag{6.22}$$

Because ds is an invariant, this expresses Newton's second law of motion so that it has the same form in any inertial reference frame. To transform from one inertial frame to another, the force is merely Lorentz transformed accordingly $\mathbf{F}' = \mathbf{M}^T\mathbf{F}\mathbf{M}$.

This leads to the nature of accelerations in spacetime. The relativistic form of Newton's second law is obviously

$$\mathbf{F} = m\mathbf{A}\,, \tag{6.23}$$

where **A** is a vector in the four dimensions of spacetime. To understand this acceleration in spacetime return to the line element with

$$ds^2 = -(cdt)^2 + dx^2 + dy^2 + dz^2 = dx^a dx_a\,, \tag{6.24}$$

where the last term implies a summation over each dx^a, $dx^1 = cdt$, $dx^2 = dx$, $dx^3 = dy$, and $dx^4 = dz$. This implied sum is referred to as the Einstein summation convention. Now divide both sides of this equation by ds^2

$$1 = U^a U_a = -U_t^2 + U_x^2\,, \tag{6.25}$$

for $U^a = dx^a/ds$ the four velocity vector. Now differentiate this by d/ds to find that

$$\frac{d}{ds}1 = 0 = 2\frac{dU^a}{ds}U_a \rightarrow A^a U_a = 0. \tag{6.26}$$

This curious result shows that in spacetime the direction of the velocity of a particle is perpendicular to the acceleration, and by Newton's second law it is perpendicular to the force. The magnitude of the acceleration is g defined by $g^2 = A^a A_a$. For acceleration along the x-axis this gives

$$U_x A_x = U_t A_t,\ g^2 = -A_t^2 + A_x^2\,. \tag{6.27}$$

Solving A_t and A_x according to U_t and U_x results in the differential equations

$$\frac{dU_t}{ds} = gU_x,\ \frac{dU_x}{ds} = gU_t\,. \tag{6.28}$$

For those unfamiliar with differential equations, these are equations built from derivatives of an unknown function, where functions are the set of solutions to the differential equation. The solutions are the functions

$$U_t = \cosh(gs),\ U_x = \sinh(gs)\,, \tag{6.29}$$

or equivalently that $t = g^{-1}\sinh(gs)$ and $x = g^{-1}\cosh(gs)$ [6.3]. The acceleration of a particle in spacetime appears in Figure 6.4. The asymptotes to the hyperbolic curves define a particle horizon. Any spacetime point to the future of the upper asymptote is unable to send a signal to the accelerating particle. This is important in future discussion on relativistic starcraft.

Special relativity and its dynamics are similar to Newton's original laws of motion. The space of Newtonian mechanics is extended to spacetime. The three laws of motion are modified accordingly.

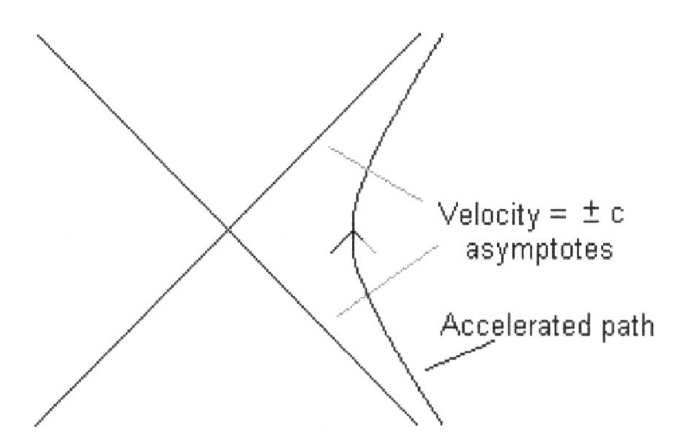

Fig. 6.4. Hyperbolic nature of acceleration in spacetime.

The first law of motion is:

- The world line of a particle in spacetime in the absence of any force is a time-like line with constant intervals of proper time along this line.

This defines a reference frame as inertial with an invariant speed of light. An appropriate observer in spacetime is on an inertial reference frame in the absence of any accelerating force. An inertial observer is able to observe the dynamics of particles and their interactions in spacetime consistent with the dynamical equation given by the second law of motion.

The second law of motion is:

- The acceleration of a particle in spacetime $A^b = dU^b/ds$, for ds an infinitesimal element of the invariant interval or proper time, with $U^b = dx^b/ds$, is proportional to a force in four dimensions F^b according to

$$F^b = m\frac{dU^b}{ds}. \tag{6.30}$$

This dynamical equation can only be properly applied for an observer on an inertial reference frame that satisfies the first law. An accelerated observer is not able to properly apply the second law of motion for particles under forces.

The third law of motion is:

- Whenever one body exerts a four dimensional force on a second body, the second body exerts a four-force with components of equal magnitude on the first, where the spatial components are in the opposite direction and with the same temporal force F^t.

This is exactly the same statement for the Newtonian case, but with the term force replaced by a four-force. Further, the reactive force is only opposite in its spatial direction. The time part, $\Delta E = dE/ds\Delta s$ term is the same on both bodies. This changes the physics so that if one body changes its momentum by $\mathbf{\Delta P} = (\Delta E,\ \mathbf{\Delta p})$ by exerting a force on a second body, the second body changes its momentum by $\mathbf{\Delta P} = (\Delta E,\ -\mathbf{\Delta p})$. This change in the third law reflects that the symmetry of spacetime is isotropic in spatial direction, as well as homogeneous, but that there is an additional transformation principle given by the Lorentz transformations, called the Lorentz group. The retention of the translation and rotation invariance seen in the standard Newton's third law with the Lorentz group defines the Poincaré group. This is now the fundamental symmetry of spacetime. It is interesting to note that for small velocities $v \ll c$ this structure recovers Newton's laws.

Just as with the standard Newton's second law the dynamics of special relativity is deterministic. Since the differential equation is second order in proper time the dynamics are completely deterministic. Given a particle subjected to a force its trajectory is completely determined into the future and into the past. The major difference is that a force can't be instantaneously imparted to a particle at a distance. Special relativity was arrived at to fix problems with electrodynamics, so that a fast moving particle in an electric field in the lab frame interacts with transformed electric field plus a magnetic field on its frame. This suggests some inconsistency with Newtonian gravity, an issue fixed by general relativity. However, special relativity recovers the property of deterministic dynamics of Newton's second law, since the dynamical equation is second order an invariant under $s \rightarrow -s$.

These dynamics will form the basis for relativistic space flight. In particular the equations for accelerations will be important. With this machinery at hand it is possible to explore the physics required for sending a spacecraft to another star.

A discussion of special relativity should include some mention of general relativity. This will not be explored in great detail, but the later discussion on exotic space propulsion methods does require some description of general relativity. What makes special relativity "special" is that it assumes that an inertial reference frame coordinate system can extend throughout spacetime. This is the case where spacetime is flat, and there is no gravitating body that curves the spacetime. If spacetime is curved a flat spacetime configuration only exists in a sufficiently small region where deviations from flatness are negligible. In general spacetime is curved, which is the cause for the adjective "general" in general relativity. This subject is vast and deep, with a massive accumulation of literature since Einstein advanced general relativity in 1916.

General relativity was motivated by the existence of gravity. Those annoying invisible lines of force in Newton's law of gravitation implied an instantaneous propagation of the gravity field. So if one gravitating body is moved by some means to another region of spacetime changing lines of gravitating force would instantaneously propagate to any other body. Yet special relativity implies that causal propagation can only be on a null or timelike path. So Newtonian gravity is inconsistent with special relativity.

A crucial ingredient in the formulation of general relativity is the equivalence principle. An inertial reference frame in flat space, with no gravity, is one where all bodies will by the first law of motion remain at rest or in a constant state of motion with respect to each other. Similarly a reference frame that is falling in a gravity field is one where bodies will remain at rest or in a constant state of motion with respect to each other. Astronauts in the space shuttle orbiting the Earth are weightless because they are falling with the shuttle and its contents at the same rate. Yet the craft is moving fast enough to keep missing the Earth which curves away from the falling path. Hence if this falling reference frame is small enough so that the radial divergence of the gravity field may be ignored the two situations, an inertial reference frame in free space and a reference frame freely falling in a gravity field, are completely equivalent. In effect the freely falling observer cancels out gravity. This equivalence principle extends to accelerated frames as well. An observer on a rocketship accelerating with a constant acceleration, say one gee, observes the same physics as an observer here on Earth. An observer on the rocket who drops a mass sees that mass drop to the floor according to $v = at$, where indeed the mass is inertial and the floor of the rocket accelerates towards the ball. Yet within the rocketship frame this observation is no different from what an observer on Earth sees with dropping a mass.

That the freely falling reference frame has to be small suggests that an inertial reference frame in a gravity field is small in extent. Gravity must in some ways involve a "patching together" of many of these local regions that can be considered as local inertial reference frames. Einstein considered Riemannian geometry, devised by Bernhard Riemann a half century earlier, as a way to approach this problem. This is a "deformation", where special relativity is a deformation of Newton's laws.

The first law is:

- The world line of a particle in spacetime in the absence of any force is a time-like curve, or geodesic, with constant intervals of proper time along this line. Since this particle moves through local Lorentz inertial reference frames the condition for zero acceleration is

$$\frac{dU^a}{ds} + \Gamma^a{}_{bc}U^bU^c = 0, \tag{6.31}$$

where the $\Gamma^a{}_{bc}U^bU^c$ is the correction term needed to glue local Lorentz inertial frames together the inertial particle passes through.

In the case that spacetime is flat $\Gamma^a{}_{bc}U^bU^c$ vanishes and special relativity is recovered. This connection coefficient is involved with the computation of spacetime curvatures. This is the geodesic equation and also defines the appropriate frames from which dynamics is studied. In the case that this connection coefficient is nonzero this equation defines the proper inertial reference frame as freely falling in a gravity field. For a weak gravity field $\Gamma^i{}_{00}U^0U^0 = -GMr^i/r^3$, for the index i over spatial dimensions, recovers Newton's law of gravitation. This indicates that the gravity force is no longer a force. The force associated with gravity is due to other forces which prevent geodesic motion of a body in curved spacetime.

The second law is:

- A body under a force will experience an acceleration defined by

$$F^a = \frac{dU^a}{ds} + \Gamma^a{}_{bc}U^bU^c\,. \tag{6.32}$$

This form of the second law of motion indicates that gravity is distinguished from any other force. In fact, gravity is not at all a force, and a system falling or behaving according the gravitation has gravity removed as a force. This is not the full version of the second law of motion. A more complete version of the second law is:

— Consider a vector V^b, which is a unit vector in the direction of a Killing vector, and its divergence $V^b{}_{;a} = V^b{}_{,a} + \Gamma^b{}_{ac}V^c$. Here $V^b{}_{,a}$ refers to the divergence or derivative of this vector in the a component direction, and reflects a deviation from geodesic or inertial motion in spacetime. Now of course this depends upon whether the spacetime admits a Killing vector. A Killing vector defines an isometry of spacetime according to how its projection onto a momentum vector is constant in the spacetime. This issue delves into subtle issues that are left open here. However, the dynamical equation of motion is given by [6.4]

$$(V^b{}_{;a}V^a)_{;b} = R_{ab}V^aV^b. \tag{6.33}$$

The term R_{ab} is the Ricci curvature term, which is a measure of how spacetime is curved. An indication of what is curvature is given below. For those acquainted with the most elementary of physics things are indeed starting to become strange! Yet this is illustrated for some small degree of education, and to indicate that this leads to the famous Einstein field equation.

The third law of motion is:

- For one mass acting upon a second mass with a momentum divergence $P^a{}_{;b}$ will experience a momentum divergence $P^{a'}{}_{;b}$ so that

$$P^a{}_{;b} + P^{a'}{}_{;b} = 0\,. \tag{6.34}$$

However, this is not the full form of this third law. The Einstein field equation is

$$R_{ab} - \frac{1}{2}Rg_{ab} = \frac{8\pi G}{c^4}T_{ab}, \tag{6.35}$$

for $R = g_{ab}R^{ab}$ the Ricci curvature scalar. The conservation law analogous to the third law of motion is that $T^{ab}{}_{;b} = 0$. The term T_{ab} is the momentum-energy tensor, which contains all the source fields for a spacetime configuration. If T_{ab} is zero spacetime can still be curved with $R_{ab} = \frac{1}{2}Rg_{ab}$, which is called an Einstein space. This will describe spacetime physics without any sources, such as gravity waves in free spacetime. Gravity waves are analogous to electromagnetic waves in a region free of charges.

This is an enormous jump in conceptual and mathematical abstraction, which is presented here for one's ability to say, "I have seen Einstein's field equation," and to see that general relativity has much the same outline as special relativity and Newtonian mechanics.

So how is curvature defined? Above we see that there is this object called the Ricci curvature, which is some measure of how a space is curved. Consider a ball with a vector at the north pole tangent to the surface there. One can do this with a pencil on a ball. First slide the pencil down the ball keeping it tangent to the sphere. Then slide the vector along the equator, without changing the pointing direction, until it has gone one fourth the way around the ball. Now slide the pencil back to the pole. This is illustrated in Figure 6.5. The vector has been rotated by an angle $\theta = \pi/2$, and the area of the sphere enclosed by this path is $A = \pi r^2/2$, or an eighth of the spherical surface area. The Ricci curvature scalar, $R = \theta/A$, is this angle

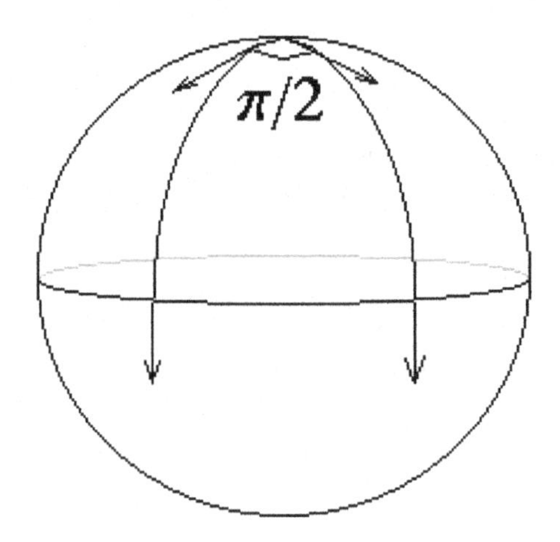

Fig. 6.5. Parallel transport of a vector on a two dimensional sphere.

divided by the area enclosed by this curve,

$$R = \frac{1}{r^2}. \tag{6.36}$$

This is the simplest definition of curvature, where in general curvature is defined by a tensor. This tensor involves components which "project" out of the area enclosed by how a vector is carried along this closed loop. A vector carried this way, called parallel translation, around a curve defines curvature components, which when summed together gives the curvature scalar defined by how its angle changes divided by the enclosed area.

One might ask, "What about quantum gravity?" Issues of quantum mechanics are largely avoided here, and quantum gravity in any depth is outside the scope of this book. However, it can be said that the pattern is clear. A successful theory of quantum gravity is likely to be an effective deformation of the three laws of motion in a manner that incorporates quantum theory in the appropriate geometric content.

This look at general relativity is meant for the reader's enlightenment and for some background for the discussion on the prospects for exotic methods of star travel in Chapter 11. This sets up some of the ideas that will be used to discuss these matters. The Einstein field equation do predict solutions such as wormholes and even warp drives. However, there are problematic issues associated with these. These solutions may in fact not really exist, but are probably mathematical artifacts.

Chapter 7

The Relativistic Rocket

The most reaching proposed method of propulsion, project Daedalus, can in principle send a spacecraft to at best about 10% the speed of light. It is clear that something more robust is required to explore a star system, particularly if it is more than 10 light years from our solar system. The specific impulse required to reach a significant percentage of the speed of light must be close to the speed of light divided by one-gee. The ideal case would be for $s = c/g \simeq 3 \times 10^7$ sec. In this case the rocket plume is composed of photons. Further, these photons must be generated by some energy source that converts half or more the initial mass of the craft into photons. This obviously requires an application of $E = mc^2$ to convert a higher percentage of mass to energy than is currently possible with nuclear energy.

Currently a direct conversion of matter to energy requires that matter be annihilated by its interaction with antimatter. Antimatter is that elixir of science fiction fame, such as the StarTrek "antimatter pods." Antimatter is a real aspect of physics, where its prediction came from Paul Dirac with his quantum equation for spin $\frac{1}{2}\hbar$ particles such as the electron. The Dirac constant $\hbar = 6.67 \times 10^{-34}$ J-sec is a quantum unit of angular momentum or action. Such particles were known to be different than particles with integer $\hbar$ spins. Wolfgang Pauli laid down how the wave functions were antisymmetric, which meant that only one of these particles, called a fermion, could exist in a single quantum state [7.1]. A consequence of this is the selection mechanism for electrons in an atom. In each atomic shell only two electrons may exist, where they do this by having oppositely aligned spins. The difference in spin states is the quantum state difference that permits these electrons to exist in the same atomic quantum state. For this reason chemistry is possible, for otherwise all electrons would drop equally into a minimal energy state. Dirac derived a relativistic version of the quantum wave equation for the electron and found that for an electron created from a "sea" of virtual quanta there was a corresponding positive hole created. This positive hole was identified as the positron or anti-electron. Its quantum numbers, charge, lepton number, etc, are opposite from that of the electron. It was experimentally found by Sanderson in 1932 that a γ-ray which impacts a heavy nuclei may cause the generation of an electron and its corresponding opposite, the positron. The photon, which has no quantum numbers for the electron or positron must create both of them so there is no net creation of these quantum numbers. Thus was born the notion of antimatter, which found its way into popularizations and science fiction in the 1960s.

Every elementary particle that exist has its antiparticle, except for the photon and some other neutral bosons. There is some debate as to whether the neutrino is its own antineutrino in the form of Majorana neutrinos. A proton has its corresponding antiproton. Indeed antihydrogen atoms with a positron bound to an anti-proton have been generated. Antiparticles are today a generic feature of elementary particle physics. So the science fiction approach to pump matter and equal proportions of antimatter into some sort of chamber appears to be the simplest "first order" engineering model for some 100% conversion of mass to energy. Of course there are some obvious problems here. A proton and antiproton will produce two photons with an energy $\simeq 1$ GeV. A method for containing these photons and directing them is an obvious complexity. An ordinary mirror simply will not work. This problem will be addressed later.

Another problem is that antimatter is not at all prevalent in the universe. Due to something called CP violations at $\simeq 1$ TeV, far below unification energy of 10^{12} TeV, the universe in its expansion shortly after the big bang resulted in an excess of particles over their antiparticles. CP symmetry means that a particle state transformed by the product of two mathematical operators to have an opposite charge as well as a parity change, or change in the handedness of a wave function, is equivalent to changing the time direction of the particle $CP = T^{-1}$. For more reading on quantum field theory see A. Zee's book [7.2]. This symmetry is slightly violated in the universe by a process that operated in the early universe. This resulted in a small excess of matter particles over their antiparticle complements. So there is no "wellspring" of antimatter. Antiparticle must be generated by imputing lots of energy into a process, such as the high energy collision of two protons with $p + p \rightarrow 3p + \bar{p}$, where $\bar{p}$ is an antiproton. Some of

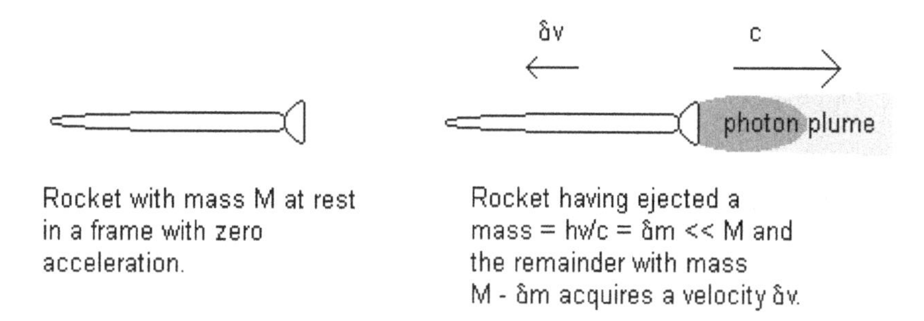

Fig. 7.1. Schematic of third law of motion concept of photon rocket.

the energy of this interaction goes into the generation of identical pairs of particles and their antiparticles, where upon the antiparticles are then magnetically bottled. This is a very expensive process that must be done with high energy accelerators. Further, at the end of the day a tiny percentage of the energy input into these machines ends up converted to antimatter. This is a high energy analogue of the "hydrogen economy," where hydrogen as a fuel must be extracted from water, by electrolysis, with the input of energy from some other source. So this is an obvious difficulty with obtaining enough antimatter required for a relativistic photon rocket.

In spite of these problems we will press on with the basic concept of the photon rocket. The issue of high energy confinement or directing energy will be discussed later in this chapter. The problem of obtaining antimatter will be discussed in the Chapter 9. We assume here that these problems have been solved. By the first law of relativistic motion the acceleration of this rocket must be measured in an inertial reference frame. The time as measured by observers on the inertial reference frame is denoted by t, called the coordinate time, and the time as measured by a clock on the accelerating rocket is the proper time and is denoted by T, previously called s. The acceleration of the rocket is denoted by g. The distance the spacecraft has travelled at any time is d, the velocity is v and $\gamma = 1/\sqrt{1-(v/c)^2}$.

The four momentum of a body in flat spacetime, such as a rocket, is

$$P = (E,\ \mathbf{p})\,. \tag{7.1}$$

The four-momentum has the spatial momentum and the energy. The energy the rocket before it starts to accelerates has the initial energy

$$E_i = (M+m)c^2\,, \tag{7.2}$$

where M is the fuel mass and m is the payload mass. Once the fuel, presumably matter plus anti-matter, is used it is converted to photon energy plus the final energy of the system

$$E_f = \gamma mc^2 + E_{ph}\,, \tag{7.3}$$

where E_{ph} is the energy of the photons generated. Conservation of energy tells us that $E_i = E_f$ and so

$$(M+m)c^2 = \gamma mc^2 + E_{ph}\,. \tag{7.4}$$

Similarly there is a conservation of momentum. Before accelerating the total spatial momentum is zero in the Earth frame,

$$P_i = 0. \tag{7.5}$$

After the fuel is converted to energy, the final spatial momentum is that of the ship plus that of the photons directed in the opposite direction

$$P_f = \gamma mv - E_{ph}/c\,. \tag{7.6}$$

By conservation of momentum $P_f = P_i$, and so

$$\gamma mv - E_{ph}/c = 0\,. \tag{7.7}$$

Eliminating E_{ph} from these two conservation equations gives

$$(M+m)c^2 2 - \gamma mc^2 2 = \gamma mvc\,, \tag{7.8}$$

and the fuel to payload ratio is then

$$M/m = \gamma(1+v/c) - 1\,. \tag{7.9}$$

Now appealing to the equations for an accelerated reference frame with

$$\gamma = \cosh(gT/c),\ \ v = c\tanh(gT/c)\,, \tag{7.10}$$

this ratio is then

$$M/m = \exp(gT/c) - 1\,. \tag{7.11}$$

So in order to accelerate to the velocity v under a constant acceleration g this is the required fuel/payload mass ratio. Now as a practical matter rockets are efficient for this ratio = 10 or so. The specific impulse is $s = c/g = 3 \times 10^7$ sec, and so

$$s\ln(M/m+1) = T = 7.2 \times 10^7 \text{ sec}\,. \tag{7.12}$$

So the rocket can accelerate for about two years at a $g = 10m/sec^2$. Put this into the equation for the gamma factor and velocity

$$\gamma = \cosh(gT/c) \simeq 12 \quad (7.13a)$$

$$v = \tanh(gT/c) \simeq .993c, \quad (7.13b)$$

which is pretty fast. For a low gamma rocket a ratio $M/m = 2$ will accelerate at $g = 10m/s^2$ for a proper time

$$s \log(M/m + 1) = T \simeq 3.3 \times 10^7 \text{ sec}, \quad (7.14)$$

which is approximately 1.05 year to reach a velocity $v = .76c$ at one gee. The Lorentz factor is $\gamma = 1.5$ that gives the coordinate time on Earth $t = (c/g)\sinh(gT/c)$ or $t = 3.88 \times 10^7$ sec or 1.23 yr, which is longer than the proper time on the craft. A velocity $v \simeq 0.8c$ is sufficient for sending a probe to a star within a radius of 50 ly.

Of course this assume a 100% efficient motor! Thermodynamics and engineering ugliness creeps into the issue. Yet it is apparent that if we could generate antimatter by some means, quantum black holes, magnetic monopoles or ... that a relativistic probe could be sent to some of the nearby stars. Also the use of rocket stages will increase this efficiency some, yet the practical limit is about 3 or maybe 4 stages. Also this does not assume the rocket decelerates to its destination. For that consider half the M/m for the midway trip and then compute the same for the return trip. For a probe only a one way trip is considered.

It is interesting to note a spacecraft that accelerates at one-gee will reach velocities very close to the speed of light relative to the Earth in a very short time. The following chart illustrates this

T	t	d	v	γ
1 year	1.19 yrs	0.56 lyrs	$0.77c$	1.58
2	3.75	2.90	0.97	3.99
5	83.7	82.7	0.99993	86.2

(7.15)

It is apparent that a clock on the continually accelerating craft will tick at a rate far slower than a clock on Earth and the craft can reach enormous distances in a time frame much smaller than time on Earth. Low gamma craft cut off at $\gamma \leq 2$ for this may be the practical limit for any probe designed to send information back to Earth within a time frame reasonable on Earth. Higher gammas do not produce significantly reduced time frames for a return message. For much higher gammas the time rate on Earth races far ahead of that on the craft. Most nation states, civilizations or empires have a life that is measured in a few centuries. A craft that is sent to higher gammas to some very distant destination will transmit its data to an increasing number of future generations here on Earth. To engage in such a program would require that humanity settle its affairs according to some long term project of sustainability over this time period. There is no historical precedent for this. In other words for higher gamma rockets it is likely that by the time the mission results are sent back in a signal back to Earth, civilization may well exist in some other paradigm, such as a dark age, where nobody is here to receive the message. Curiously the seeds of a dark age may already be germinating in our age as seen with resource depletion issues and the current rise in the social interest and power of religious beliefs.

It may be possible to send crews on such higher gamma craft, but the information they garner may be held only by them. For exceedingly high gammas this is almost certainly the case. If such craft could be arranged the crew could reach space beyond our galaxy with 11 years of their proper time, and reach the observed boundary of the entire universe within about 25 years of ship time. Yet what ever knowledge they obtain is theirs, for by this time conditions on Earth will have evolved beyond anything predictable, and in the latter case Earth will have been toasted by death of the nova stage of the sun. The universe when it comes to such space travel is similar to a black hole: Those who wish to travel to such heights will exit permanently the society they leave behind. Of course such a space mission could only be prepared by some exceedingly wealth class of "lunarians," who have technologies far beyond those considered here, which are in turn technologies beyond today's. Such craft involve ideas such as the Bussard ramjet that can consume vast amounts of material coming in front of the craft. Such architectures are outside any consideration that could ever be executed by any reasonable plenipotentiary structure.

It is most likely that a photon rocket will have an acceleration less than one-gee [7.3]. For a proper time T the value of γ is $\gamma = \cosh(gT/c)$ and for $T = (c/g)a\cosh(\gamma)$ then it takes a proper time $T = 3.95 \times 10^8$ sec or 12.5 years for $g = 1$ m/sec^2 to reach $\gamma = 2$, and $T = 1.98 \times 10^8$ sec or 6.3 yr for $g = 2$ m/sec^2 to reach the same velocity. The coordinate time as measured here on Earth is $t = (c/g)\sinh(gT/c)$ are then 5.2×10^8 sec or 16 years for $g = 1$ m/sec^2 and about 8 years for $g = 2$ m/sec^2. The distance travelled may be shown to be

$$d = (c^2/g)(\cosh(gT/c) - 1), \tag{7.16}$$

which is 9×10^{16} m or about 9.5 light years for $g = 1$ m/sec^2 and about half that for $g = 2$ m/sec^2. A computer run of various accelerated missions that reach $\gamma = 2$ are seen in equation 7.16 below. A spacecraft accelerated to $\gamma = 2$ sent on a mission 20 light years distant a 1 m/s^2 craft will take $\simeq 33$ years of coordinate time to reach the destination, while a 10 m/sec^2 craft will take 21.3 years to reach the destination. This shows a rocket with a higher acceleration does not produce great savings over a lower accelerated rocket. For very high accelerations to $\gamma = 2$ the limiting time is obviously 20 years. High gamma rocket produces less and less mission time as measured on Earth compared to low gammas.

g (m/s^2)	T (yr)	t (yr)	d (ly)
1.0	12.54	16.49	9.525
2.0	6.271	8.248	4.762
3.0	4.181	5.499	3.175
4.0	3.136	4.124	2.381
5.0	2.509	3.299	1.905
6.0	2.090	2.749	1.587
7.0	1.792	2.356	1.360
8.0	1.567	2.062	1.191
9.0	1.394	1.833	1.058
10.	1.255	1.650	0.952

(7.17)

These results illustrate the length contraction and time dilation of relativity. An observer watching a rocket accelerate away will find that the time measured on the rocket's clock is slowing down. From the perspective of an observer on the rocket the time required to travel a distance is smaller than what her compatriots on Earth observe. Similarly, the distance between two points along the direction of motion is reduced according to the astronaut. This length contraction means that the time required to traverse the distance between two points is reduced. For a $\gamma = 2$ relativistic rocket this distance is reduced by $\frac{1}{2}$, which means that the corresponding proper time required to travel that distance is also halved. From the perspective of an observer on Earth there is no change in the distance between two points, but the rocketship is observed to be shortened by half and its clock is seen to run at half speed. Figure 7.2 illustrates how an observer on a

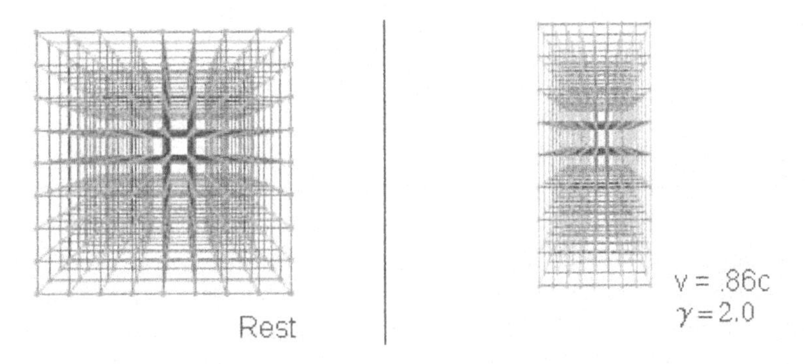

Fig. 7.2. The length contraction of a lattice of points seen by a relativistic observer.

rocket with $\gamma = 2$ will see a lattice of points in space. This illustrates the "clocks and rods" view of relativity. An observer who is accelerating will see the distance between these lattice points continue to shrink. The distance between lattice points asymptotically approaches zero as the velocity $v \rightarrow c$.

A minor consideration needs to be addressed as well. The rocket travelling through space at a $\gamma = 2$ had best not run into a small particle. Even a microgram mass particle that runs into the spacecraft will have a devastating impact. The energy released would be $\simeq mc^2$, which would be $9 \times 10^7 J$, which is equivalent to about $40lbs$ of explosives. Even ordinary hydrogen in interstellar space would cause damage, as it would become a radiation flux on the ship. Thus the craft would be best made needle shaped. A radiation field in front of the ship might be used to ionize the hydrogen and a magnetic field used to deflect the charged particles. Potentially some form of vaporizing beam needs to be employed in front as well to eliminate threats due to larger particles such as dust.

This is the basic theory for the relativistic photon rocket. It is apparent that there are some technical problems here. The first big one is how are photons with energies nearly a billion times the energy of optical photons to be harnessed or channelled into a thrust? A large flux of these photons will destroy any material they run into. Obviously something must absorb these photons or scatter them into lower energy photons. If the matter-antimatter interactions involve electrons and positron an obvious approach would be to Compton scatter these photons. Here the photons scatter off of charged particles to yield some of their energy to this particle and so the photon has a lower energy. This might in a rough way be extended to proton-antiproton interactions as well, even though this ignores processes such as the production of π mesons.

The four momentum of a photon is $hc\nu(c,\ \mathbf{n})$, where h is Planck's unit of action in quantum mechanics, ν is the frequency of the photon, and $\mathbf{n}$ is the direction the photon propagates in space. The four momentum of a particle of mass m is $(E,\ m\mathbf{v})$. The momentum of the photon and particle is conserved in the scattering process. The energy of the photon and particle changes by

$$h\nu + E = h\nu' + E', \tag{7.18}$$

and similarly the photon and particle momentum change by

$$hc\nu\mathbf{n} + m\mathbf{v} = hc\nu'\mathbf{n}' + m\mathbf{v}'. \tag{7.19}$$

For simplicity the initial momentum of the charged particle is considered to be zero $m\mathbf{v} = 0$. By squaring the momentum of the electron after the scattering with $\mathbf{n} \cdot \mathbf{n}' = \cos(\theta)$ the change in the wavelength of the photon is shown to be

$$\lambda' - \lambda = \frac{h}{mc}\big(1 - \cos(\theta)\big). \tag{7.20}$$

The term h/mc is the Compton wavelength of the particle, which for the electron is 3.86×10^{-11} cm. The wavelength of radiation produced by $e - e^{+}$ annihilations is $\simeq 10^{-10}$ cm, and so an ionized gas with electrons will multiply scatter these photons. If these electrons are dense enough multiple scattering will increase their wavelength by a factor of around 10^{3}. Similarly a gas of nucleons will scatter $\simeq$GeV photons to the $\simeq 10$ MeV range in energy, which is considerably more modest.

A possible way to contain the energy of an matter-antimatter interaction is to consider the following reaction

$${}_2\mathrm{He}^4 + \bar{p} \rightarrow T + \gamma. \tag{7.21}$$

In this way a substantial amount of the energy of the interaction is carried off by the tritium ion. If we assume that the initial four momenta of the ${}_2\mathrm{He}^4$ and $\bar{p}$ are considered to have zero spatial momentum then

$$E_T + E_\gamma = \gamma mc^2 + h\nu = Mc^2, \tag{7.22}$$

and

$$h\nu/c = \gamma mv, \tag{7.23}$$

for M the mass of the initial He + $\bar{p}$ and m the mass of the tritium nucleon. It is easy to show that

$$\gamma(1+\sqrt{1-\gamma^2}) = M/m, \tag{7.24}$$

or

$$\gamma = \frac{m}{2M}\left(1+\frac{M^2}{m^2}\right), \tag{7.25}$$

which gives a $\gamma = 1.13$ ($v = .47c$) and the tritium has a kinetic energy $K = (\gamma - 1)mc^2$ or $.4m_pc^2$. This is 8% conversion of the initial mass energy into the kinetic energy of a charged nuclei. There are still photons with an energy of 1.61 GeV, which are very high energy.

Similarly consider the reaction

$$n + \bar{p} \rightarrow e^- + \gamma. \tag{7.26}$$

Much of the kinetic energy is transmitted to the electron. For the electron $\gamma \simeq 1000$ and the energy of the gamma ray photon is $\simeq 1.5$ GeV. This is a 25% conversion of the mass-energy into the kinetic energy of a charged particle. So there is still considerable energy not given to the photon, where it is evident that only for $\gamma \rightarrow \infty$, for $m/M \rightarrow 0$ do we still get half the energy output in a photon with a 1.0 GeV energy. There is a further problem that a neutron and an antiproton have no electrical attraction for each other and so do not readily combine.

So consider this process as something that produces a gas of photons with some energy with some reaction mass particles. With the ${}_2\mathrm{He}^4 + \bar{p} \rightarrow T + \gamma$ process there would then exist a gas of photons and tritium nuclei. Under multiple scatter of the nuclei by these photons there will exist a thermodynamic equilibrium. In other words the energy of the photons will become distributed equally amongst the tritium nucleons. This is the equipartition theorem of statistical mechanics. So the tritium nuclei with a mass-energy equivalent of 3 GeV will come to equilibrium with an energy of 1.61 GeV. This means that these nucleons have an energy that is 54% of their rest mass. The temperature of this gas of tritium nucleons would be around $\simeq 10^{12}$ K. The average velocity of the outward plume would then be $v \simeq 0.58c$. Some process must permit the existence of this plasma, so that the actual rocket material does not come in contact with energies and temperatures at this level. Obviously this plasma must be magnetically contained. The problems here are quite evidently manifold.

From here the pure photon rocket has been modified into a rocket with a high velocity plume of massive particles. A rocket with a four momentum $P = (mc^2,\ 0)$ that shoots out some δm of particles with a velocity u will then have the four momentum $P' = (\delta\gamma(m - \delta m)c^2,\ \delta(\gamma v)(m - \delta m)c)$, $\gamma = \gamma(v)$ due to the plume four momentum $\delta P = (\gamma'\delta mc^2,\ -\gamma'\delta uc)$, with $\gamma' = \gamma'(u)$, so that

$$mc^2 = m\delta\gamma c^2 + \gamma'\delta mc^2 m\delta(\gamma v) = \gamma'\delta mu. \tag{7.27}$$

The momentum part gives the differential expression

$$\gamma\delta v + v\delta\gamma = \gamma' u\frac{\delta m}{m}, \tag{7.28}$$

where it is easier to convert this to an integral in $d\gamma$ with

$$\int_1^{\gamma_f} \frac{d\gamma}{\sqrt{1-\gamma^{-2}}} = \gamma' u \int_M^m d\frac{m'}{m'}, \tag{7.29}$$

with the result

$$\gamma\sqrt{1-\gamma^{-2}} = \gamma' u ln(M/m). \tag{7.30}$$

The gammas and mass ratios of interest are displayed in the chart below

γ	M/m
1.5	1.674
1.6	1.779
1.7	1.885 .
1.8	1.994
1.9	2.106
2.0	2.222

(7.31)

These gammas and mass ratios are comparable to pure photon rocket. The accelerations, distances and times of flight for the pure photon rocket derived above hold for this modified version as well.

A spacecraft that annihilates matter is commonly thought to use antimatter, as indicated above. This antimatter is then contained in "antimatter pods," to use the term from *Star Trek*, which are essentially tanks or bottles. Clearly this antimatter must not touch the walls of any container, for if it does an explosion will result. In the case of antiprotons, which are negatively charged particles, a magnetic field can bottle these particles as the moving charged particles spiral around the magnetic field lines. If the magnetic field pinches off at the ends of the tanks this acts as a sort of mirror. However, magnetic fields do not pinch off completely, so the ends will be leaky. Further for $\gamma = 2$ it is clear that the mass ratio is larger than 2. This requires bottling up lots of antiprotons. The mutual electrostatic repulsion between these particles will be enormous and impossible to hold. Thus a spacecraft that consists of $\simeq 1/4$ antimatter will have to contain it in a neutral form, such as anti-hydrogen — an antiproton with a positron around it. However, this is neutrally charged and a magnetic bottle will not work. It will be hard to bottle up anti-hydrogen so that it does not touch the walls of a container.

In quantum field theory there is a conserved quantity called the baryon number B. A proton has a baryon number $B = 1$ and an antiproton has $B = -1$. Thus a process $p + \bar{p} \rightarrow 2\gamma$ will conserve a net zero baryon number. Baryon number, which is based on more fundamental quantum numbers associated with quarks, appears to be a strictly conserved, and no experiment has observed any violation of B. However, black holes will violate baryon number, and some theoretical physics with Grand Unified Theories (GUTs) indicate possible violations of B as well. So it is possible that with GUT physics or with quantum black holes that matter might be directly converted to energy by violating B. These prospects will be discussed in Chapter 9 on the technical requirements for starcraft.

The next approach for star travel to be discussed here is the photon sail. This approach does not suffer from some of the technical difficulties of the relativistic rocket. So it would appear that the photon sail is more likely to first travel to a star. However, the photon sail requires delicate construction in space on a colossal scale. This poses difficulties for the photon sail that might be just as daunting as those with the relativistic rocket. If the technical difficulties discussed above can be solved the relativistic rocket would be much more modest in size, maybe not any larger than a standard launch vehicle.

Chapter 8

The Photon Sail

The photon exerts a pressure on a surface. It is not a large pressure in the case of solar radiation, but is present as seen with the Nichols radiometer. This pressure $P = 4.6 \times 10^{-6}$ p for solar radiation on the Earth's surface requires large amounts of surface area for it to create a significant force. A solar sail with a radius R would then experience a force $F = 2\pi R^2 P$. For a solar sail of 100 km in radius the force on the sail would be about 3×10^5 N. For a sail material made of aluminum 16 nanometers thick the volume would be 1000 m^3 or a billion cubic centimeters. Since the density of 2.7 g/cm^3 this is 2.7×10^7 kg of material. Thus the acceleration would be $g \simeq .011$ m/sec^2. This estimate involves the absorption of the photon, where the reflection of photons would double this acceleration. The small acceleration will create a longer duration for the time of flight, but it is reasonable to consider a photon sail can reach low gammas. A version of the interstellar photon sail was studied by Robert Forward, with the proposed *Starwisp* craft [8.1].

The same relativistic analysis is employed to solve this problem. The energy the spacecraft before it starts to accelerate has the initial energy

$$E_i = mc^2 \,. \tag{8.1}$$

The solar sail converts some of the photon energy to kinetic energy. For E_{ph} the photons sent to the craft and E_{ref} the photon energy reflected back and the final energy of the system is

$$E_f = E_{ph} + mc^2 = \gamma mc^2 + E_{ref} \,. \tag{8.2}$$

Similarly, momentum is conserved. Before accelerating the total spatial momentum is zero in the Earth frame.

$$P_i = 0 \,. \tag{8.3}$$

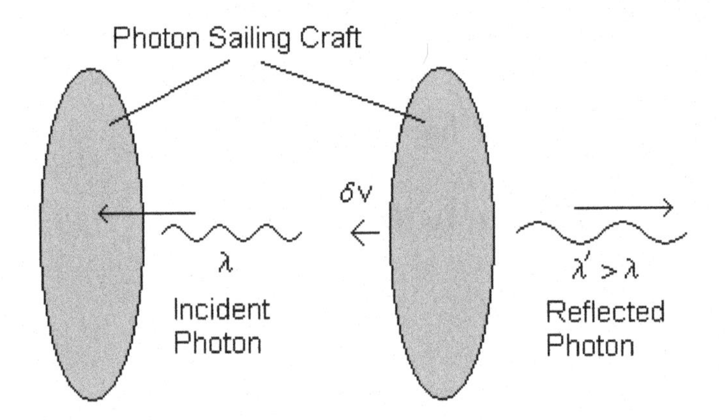

Fig. 8.1. Schematic of momentum transfer from photons to a light sailing craft.

The final momentum is the momentum of the solar photons incident on the sail, which is equal to the momentum of the spacecraft plus that of the reflected photons is

$$P_f = E_{ph}/c = \gamma mv - E_{ref}/c\,. \tag{8.4}$$

The energy of the reflected photons is related to the energy of the spacecraft by

$$2E_{ph} = \gamma mc(c+v) - mc^2\,. \tag{8.5}$$

Since $v^2 = c^2(1 - 1/\gamma^{-2})$

$$2E_{ph} = \gamma mc^2(1 + \sqrt{1 - 1/\gamma^{-2}}) - mc^2\,. \tag{8.6}$$

Below is a table that compares gammas with the fraction E_{ph}/mc^2

γ	E_{ph}/mc^2
1.5	0.559
1.6	0.624
1.7	0.687
1.8	0.748
1.9	0.808
2.0	0.866

(8.7)

Equations 8.5 and 8.6 indicate the need to reflect a lot of photons! Assume that this sail craft is powered by some solar collector near Earth. If the collector is near Earth there are about 1500 watts/m^2 of power of solar radiation. For a year as $\sim 3\times 10^7$ sec a meter would collect 4.5×10^{10} J. So the mc^2 equivalent is about 5×10^{-7} kg. So collecting area is important. To reach a $\gamma = 2$ in one year requires an area of around 1.7×10^6 m^2 to accelerate 1 kg to this speed. The collector would require a radius of 525 meters, where this assume 100% conversion of solar light energy to a beam that drives the spacecraft. Clearly for a larger spacecraft the area goes up proportionately. For the 10^7 kg craft mentioned above this would imply a light collecting area with a $\simeq 1500$ km radius.

The photon sail offers the advantage of not requiring exotic engineering and physics. It also clearly has potential performance capabilities of the photon rocket. However, it requires the construction of large solar collectors. Clearly as this photon sail clears the solar system it will have to be driven forward by a directed beam of light energy. This will require some sort of solar concentrator or a solar driven laser beam of truly huge proportions. A 500 m radius collector receives 2.35 gigawatts of power, where in reality the conversion losses will require a $\sim$ 1600 m radius collector. Hence the scale of things becomes colossal and requires large scale fabrication in space.

An examination of the times and distances required to accelerate to a $\gamma = 1.15$, or $v \simeq .5c$, for an average acceleration $g = .01$ m/sec^2 to $g = .1$ m/sec^2

g (m/s^2)	T (yr)	t (yr)	d (lyr)
0.01	515.3	540.85	142.9
0.02	257.7	270.4	71.43
0.03	171.8	180.3	47.62
0.04	128.8	135.2	35.71
0.05	103.1	108.17	28.57
0.06	85.89	90.14	23.81
0.07	73.62	77.26	20.41
0.08	64.42	67.61	17.86
0.09	57.26	60.09	15.87
0.10	51.53	54.08	14.29

. (8.8)

The time and distance performance is not very good, but a solar sail with the higher end of these specification may well be able to reach some of the nearer stars within a 5–10 ly distance. This performance might be improved if the light is further concentrated. Assume that the average accelerations are increased by a factor of 2.5. The performance is then

g	T	t	d
0.025	206.1	216.3	57.14
0.050	103.1	108.2	28.57
0.075	68.71	72.11	19.05
0.100	51.53	54.08	14.29
0.125	41.23	43.27	11.43
0.150	34.36	36.06	9.524
0.175	29.45	30.91	8.163
0.200	25.77	27.04	7.143
0.225	22.90	24.04	6.349
0.250	20.61	21.63	5.714

. (8.9)

This craft for $g = .12\ \mathrm{m/s^2}$ could reach nearby stars within $\simeq 12$ ly within a 44 year time period and data could be received on Earth within 56 years. This is within the acceptable range of mission times. The star α Centuri is 4.36 ly, and τ Ceti is 11.8 ly from Earth. These are the two G-class stars similar to the sun that would be of value exploring. There are in addition 25 other stars, mostly M-class red dwarf type stars, within this range. The star ϵ Eridani, a K-type star 10.4 ly away has two planets, so this star system is a candidate for direct exploration just barely within these performance parameters of a starsail.

Above it was mentioned that these were average accelerations. Given a photon source, whether as collimated solar radiation or a laser, from the perspective of an observer on the sail craft the photons will reach the craft increasingly red shifted. As such the photon pressure will decline. An acceleration of a photon sail could only be made constant if the flux is increased in a manner to compensate for this. To see how the acceleration of a photon sailcraft depends upon its velocity a simple relativistic analysis is required. Consider a mass with a velocity $\mathbf{v}$ which reflects a photon send from Earth with a frequency ν to a photon with frequency ν' and attains a velocity $\mathbf{v}'$. Conservation of relativistic momentum gives the relationships between the spatial momentum and energy as

$$(\gamma' v' - \gamma v) = \frac{h}{mc}(\nu' + \nu)(\gamma' - \gamma) = \frac{h}{mc^2}(\nu' - \nu), \qquad (8.10)$$

where h is the Planck unit of action. Eliminating the reflected photon frequency these give the following,

$$\gamma'(c - v') - \gamma(c - v) = -\frac{2h}{mc}\nu. \tag{8.11}$$

A differential expression for the velocity may be obtained for a variable n as the number of photons incident on the craft on it frame,

$$\left((c - v)\gamma^3\frac{v}{c^2} - \gamma\right)dv = -\frac{2h}{mc}\nu dn. \tag{8.12}$$

The rate at which the photons are incident on the craft is related to the rate photons are sent from Earth by $dn = \gamma^{-1}dN$. For photons sent at a rate $R = dN/dt$ this leads to the differential equation

$$\frac{dv}{dt} = \frac{2hR}{mc}\left(\gamma^2 - (c - v)\gamma^4\frac{v}{c^2}\right)^{-1}. \tag{8.13}$$

This does predict that the acceleration will drop to zero as the velocity approaches the speed of light. Yet to understand the dynamics of a photon sail this differential equation must be integrated. This is done numerically with the result indicated in Figure 8.2. It is clear that the photon sail is best for $\gamma \le 1.25$ or for $v \le 0.6c$, which is sufficient for exploration of the interstellar neighborhood. In this case the acceleration declines to $\simeq .42$ the initial acceleration and the average acceleration is $\simeq .75$ the initial acceleration. These will decline further over time if the sail craft is sent further.

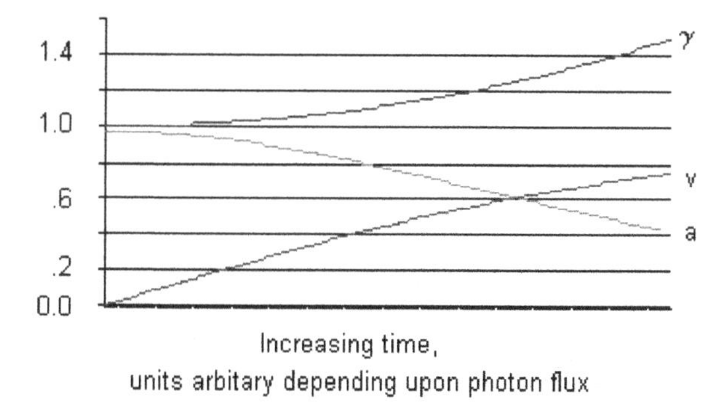

Fig. 8.2. Evolution of the velocity and acceleration of a photon sail.

For a $\gamma = 1.15$ or $v = .494c$ the acceleration declines to .726 the initial acceleration and has an average of .882 its initial acceleration. A photon sail craft which reaches half light speed is capable of reaching stars within the 20 light year range within an acceptable time frame.

The simplest way to concentrate light is with a lens. A lens with a focal length equal to the distance between it and the sun will cause the diverging light to become collimated into a beam. A plano convex lens with a radius of $r = 1500$ km situated 1.5×10^8 km from the sun will then have a radius of curvature as measured from the enter of the sun given by the thin lens equation

$$R = (n - 1)\mathcal{F}, \tag{8.14}$$

which is very large. The angle of the radial vector along the lens would be $\theta \simeq 2$ r/R. which for $n = 1.5$ it is easily found this lens will have a thickness of ~ 60 m. Obviously this is not practical due to the huge mass of such a lens. However, a Fresnel lens will work. The material set into concentric Fresnel zones which act as a slice of the lens at some radius. The sum of these over all concentric zones will concentrate light in the same fashion that a single lens does [8.2]. This technique is used to direct large beams of light from lighthouses. Fresnel lenses can be stamped onto thin sheets of plastic, such as seen with the overhead projector. It should be possible to position a huge Fresnel lens in the environs of the inner solar system and to direct a collimated beam of solar radiation.

Large sail structures have to be masted, just as sails were masted on sailing ships. Without this slight fluctuations in photon pressure on these large areas will induce wave motion and cause the structure to flap and lose its shape. This requires of course added mass to the structure. This would generically consist of a large boom protruding from the center of the back face of the sail. From there guy wires attach to points on the sail. The photon pressure pushes the sail forward, where tension on these lines preserves the shape of the sail. This may be augmented by rotating the whole sail to maintain a baseline tension on the guy wires. For the sail ship this affair can be relatively modest, for in principle it can maintain a constant direction. So there is no need for any steering mechanism required. This is less the case for the Fresnel lens. It must maintain a photon beam at an angle that has wiggle tolerance of 1×10^{-13} rad, or the beam will start to drift away from the target photon sail. Hence the directionality of this huge structure is a critical issue. By shifting the center of mass of the craft from the center of area will induce a torque on the Fresnel lens. Yet this will be slow and sluggish. A more advanced approach might be to have the Fresnel zones adjust their geometry. This would require that the whole lens contain a matrix of "smart materials" and nanotechnologies that can adjust the optical properties or directionality of the lens.

Another difficulty with a huge Fresnel lens is that it is in orbit around the sun. This compounds the steering problem enormously. A further problem is that the sun's rays will only be normal to the lens for one time out of the year. Even with smart materials that adjust the geometry of the Fresnel lens there will be a narrowed period of time where this could be used to directly propel the spacecraft. So it appears that a direct use of a Fresnel lens to propel the craft is not practical. However, the Fresnel lens could still be used to power a solar sail indirectly. It could be designed to concentrate solar radiation on a power station on the moon, which could be part of a cis-lunar solar power generation grid [8.3]. If the lens were placed at the L1 lagrange point it would be comparatively stationary and could direct solar radiation to a power station at one of the lunar poles. Mirrors in lunar orbit might be needed to redirect this light to the right point on the lunar surface. This would then power a huge laser that directs photons at the starsail. Clearly care is needed to insure against steering the Fresnel lens towards the Earth.

A 100 km radius Fresnel lens could deliver around 1.0×10^{14} w of power to this power station. Recall that this amount of solar energy could sent a 2.7×10^7 kg craft with an acceleration $g \simeq .011$ m/sec^2. With the Fresnel lens this power could be converted to laser light, where we assume a 20% efficiency and directed to a 10 km solar sail with a mass of $\sim 10^5$ kg. The acceleration is then $g \simeq .22$ m/sec^2. This gives a performance that is somewhat more in line with what is desired. Directing photons from lasers fixed to the surface of the moon would cure the small vibration problems inherent with attempting to direct photons many light year distances to a structure in space.

The problems with photon sailing are nearly as daunting as the relativistic rocket. This involves large scale construction in space, which is considerably beyond current capabilities. It also appears to require the construction of a power station on the moon which is able to generate power at a rate 1 to 2 orders of magnitude larger than all the energy currently generated on Earth. This power station would likely be as large and complex as the Clavius base depicted in the movie *2001 A Space Odyssey.* Such things certainly do not come cheap.

Robert Forward proposed a photon propelled craft called Star Wisp. The sail craft is to be propelled by microwaves generated by a large solar power station in space. The sail consists of a web or mesh of microwave

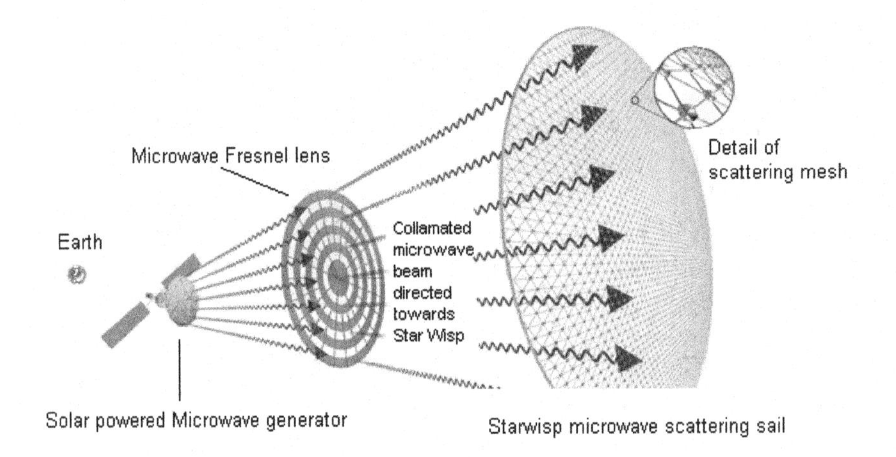

Fig. 8.3. Diagram of the proposed Star Wisp sail craft.

reflecting elements controlled by computer chips or smart materials. This sail is proposed to have a mass of 16 grams with a radius of 3 km. The microwaves are collimated into a beam by a microwave analogue of a Fresnel lens. This lens consists of rings of aluminum which act as Fresnel zones. Just as an optical Fresnel lens in a watchtower concentrate light in a direction this lens directs microwaves towards the sail craft. The Star Wisp is proposed to be accelerate at 115 gees (1130 m/s^2), and is thought to only reach a fifth the speed of light, or $\gamma = 1.02$. This limitation is due to the long wavelength of microwaves, which become redshifted relative to the frame of the craft, and that such radiation has lobes which make tight focusing difficult.

The starwisp sail would consist of a mesh of wires spaced apart a distance equal to the wavelength of the microwave radiation. This would mean that virtually all of the microwave radiation would interact with the sail, ideally with 100% of this radiation reflecting off. The dimensions of this craft are then comparatively modest. The Starwisp could in principle be constructed within a couple of decades. Robert Forward envisioned the sail as superconducting which is a perfect mirror. This assumption was overly optimistic. The mesh will absorb some of this radiation, and keeping it in a superconducting state might prove to be too difficult.

The microwires in this mesh are proposed to hold nanotechnology and nano-computer circuitry. Thus the whole sail will act as an integrated data collection system and data transmission system. These systems also are designed to adjust the tension in the sail at all times in order to keep it from responding to slight differences in photon pressure. Since the operations of the craft are distributed across the entire sail there is no rigging, or static strutting of the system. The sail would be an integrated network or neural-like net of processors which can perform all the data collection required and to maintain the trim of the sail as it is being accelerated. Of course at this time such technology does not exist completely.

The craft would have to be constructed in a lithographic manner. The wires and processors would have to be etched into some matrix of material. This material is either folded into a bundle and unfurled, or stored in pieces which are assembled in space. The sail would be contained in some epitaxial matrix which would then have to be removed. This could be some form of UV sensitive plastic which dissolves after exposure to solar radiation or some other form of evaporating material.

Space based conversion of solar radiation to microwaves has been proposed since the 1970s as a possible way to convert large amounts of solar radiation into electrical power. A 10 gigawatt microwave power generator is then proposed to be the energy source for the Starwisp. Ultra-thin and very long filaments of aluminum in a large series of concentric circles then compose a Fresnel lens that focuses this energy onto the sail. This Fresnel lens is expected to be 25,000 km in radius. Constructing this lens would be a formidable challenge. If it is constructed of ultralight materials it will still end up having a mass of 6×10^4 metric tons. Due to the limited range of microwave systems the Starwisp is propelled quickly to $.2c$, where it then cruises to its interstellar destination. The microwave beam would then be activated again on the craft as it flies by its target. The microwaves will be too weak to provide any propulsion, or braking thrust, but it could be used to power the systems on the craft.

The microwave power satellite would only need to propel the craft for several days. Thus a star system could be visited by a flotilla of such craft which probe particular regions of the stellar system. Such flyby missions would not be able to probe in detail the surface of planets in a star system, but as with planetary explorations the first missions to planets we later landed probes on have been flyby missions. The Starwisp could be used to explore planetary systems within about 10 lightyears. The missions could be accomplished within a program to place solar power stations in geosynchronous orbit.

The largest difficulty with the Starwisp is in maintaining the trim of the sail. A slight deviation in microwave pressure on one part of the sail could cause it to start flapping or twisting up. The beam profile of lasers and masers are not 100% uniform, where there are small deviations in the distribution of power through the beam. A very small deviation in the pressure on the sail could lead to a small deformation of its shape, which would then amplify the asymmetry of pressure on the sail as deformed regions reflect the radiation in non-uniform directions. This clearly requires that the tiny wires be dynamically held in place against such small perturbations. Another problem with such a light weight craft would be the presence of interstellar hydrogen. This will cause damage to nanoscale structures over time. Forward's Starwisp envisions no shielding against this assault, where obviously shielding will increase the mass of the craft considerably.

The Starwisp is a candidate for the first photon sail craft to be sent to the stars. It has the advantage of being light weight, far less massive than the optical photon craft considered above. However, it may leave little in the way of much detection and measurement apparatus on board. A craft that reaches another solar system will do so in order to make measurements and to perform astronomical observations of planets. In the case of a photon sail craft it will unlikely be able to visit every planet in the system. So it will have to take a decent telescope to observe the system. Weighting in at 16 grams the Starwisp could only at best give us a tantalizing glimpse of an extrasolar system. It is likely that the Starwisp would be scaled up sufficiently to carry the appropriate instrumentation to conduct research. Even with micro-miniaturization to electronics it is not possible to get around the need for mirror sizes and the mass required to conduct a reasonable scientific survey.

It is likely that the starwisp is the extreme light mass end of the photon sail, while the larger mass optical photon sail is at the large mass end of the scale. It is then likely a compromise will be arrived at so that photon driven starsail craft with a mass of $\sim .1$–10 tons will emerge as the realistic craft of choice with $\gamma \simeq 1.2$.

A starsail must by some means come to a halt as it approaches its target star [8.4]. It is likely that more advanced starsail craft will not just perform a quick flyby mission which would last a few days. The simplest way to accomplish this is to have the disk sail contain an inner disk that detaches from the main disk. Photons continue to reach the larger annulus. These photons reflect off of it and reach the smaller disk. This will then act to break the motion of the smaller craft relative to the star. The annulus will have to adjust its shape slightly to focus the photons onto the decelerating craft. This presumably could be accomplished with guy wires. The annulus

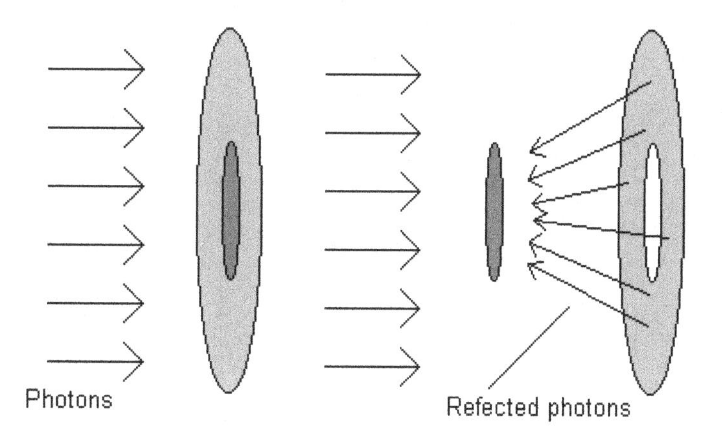

Fig. 8.4. The braking process for a photon sail.

will accelerate away as an expendable part of the craft, while the smaller central disk will slow down sufficiently to begin its exploration of the stellar system. The main sail, the annulus, would then accelerate onward or be sent into the star so as not to be some relativistic bullet flying through the galaxy. The sail craft would then use the light from the star to navigate around the star system.

If interstellar probes are indeed launched it is likely the photon sail will be the first to reach other stars. There are only about 10 stars that are accessible by this method. The relativistic rocket has about 100 stars accessible to it, as defined by the ability to deliver a message back to Earth within 50 years. It is possible that an energy architecture will involve converting solar energy collected from space into a form that can be used on earth. A lunar power station that accumulates concentrated energy from large Fresnel lenses might divert some of this energy to push photosail spacecraft to the stars [8.3]. Whether any of this comes to fruition is a future's guess. The possible impetus for this will be energy short falls here on Earth, which might in the future push our energy grid into the cis-lunar region. If this future energy pathway is able to produce sufficient energy for requirements here on Earth a possible excess might be used to propel spacecraft to the stars.

Chapter 9

Scientific and Technical Requirements

It is evident that technology of the early 21st century is inadequate for either the relativistic rocket or the starsail. The starsail is closer to current capabilities in principle. Yet this will require spaceflight capabilities and infrastructure far greater than what currently exists. The relativistic rocket does not require massive space infrastructure, but it appeals to exotic physics and engineering that do not exist. In either case the challenges are very daunting. This is a very brief overview of some of the issues that have to be resolved and some potential ways they might be handled.

For a photon sail approach the Fresnel lens, the sailcraft and a space based or lunar based power station must be constructed or assembled in space. This will require many thousands of tons of material for the lens, power station and sailcraft. This is certainly possible in principle, but beyond current abilities. The space shuttle has lofted large payloads into orbit, such as the Hubble Space Telescope and the modules for the ISS spacestation. The photon sail and its space system requirements reflect three or more orders of magnitude increase in mass scale. Further, this has to be accomplished at Lagrange points or in the cis-lunar environment. Hence the space construction problems are much larger than anything accomplished with the space shuttle. Further, a power station must be constructed that is capable to generating up to about 10 times all the power currently generated on Earth. Further, this power is to be used by a huge laser. Further, if this is based on the moon it implies lunar activities on a large scale. So far no human has stepped on the moon since December 1972.

Back in the halcyon days of the space program such infrastructure appeared likely. The space shuttle was advanced as the work horse that would facilitate such space activities. Yet the history of the shuttle has fallen far short of these early expectation. It was advanced as a platform for the launching of satellites, but was not used nearly as extensively as envisioned and the last applied satellite system was deployed in 1993, and the Chandra satellite was the last astronomic system deployed in orbit. Since then the shuttle has not lofted a satellite. Its four missions for the Hubble Space Telescope appeared to have been worth the shuttle flights. While the construction of the ISS spacestation has been a success, this success appears to be one without a real purpose or mission. The crash of the Colombia in 2003 has effectively stranded the ISS. There are current plans for a lunar base, but this plan appears so far to be malformed with no real purpose. This lunar base, if it is established, will literally be a pup-tent compared to what would be required to furnish a power station and laser capable to pushing a sail craft to the stars.

It might be that the recent new space initiative [9.1] will propel developments required to eventually generate the infrastructure required for a starsail. First off it requires that any manned spaceflight and basing of personnel on the moon have some credible scientific purpose. This would mean that lunar missions facilitate astronomical facilities, such as gravity wave detectors and optical interferometers, where astronauts will deploy and maintain equipment in ways that robots or telepresence from Earth cannot. This implies at first intermittent missions to the moon instead of a permanent human presence on the moon. Whether this will lead to a permanent human presence on the moon is uncertain. The problem is that robots and telepresent capabilities are likely to far outpace any advancement in manned spaceflight. This might mean an expanded robotic presence on the moon a lunar powerstation might be constructed and maintained robotically.

The rocket equation is a major impediment for expanded space flight. The fuel and energy requirements are very large. Further, at the early 21st century energy depletion issues are starting to make their presence known. A highly speculative way to solve this problem has been suggested, it is the space elevator. This is a very massive construction that consists of a cable attached to some point on the Earth's surface and rises up past the radius $R = 37,000$ km for geosynchronous orbit. Beyond this point is a large mass. That this mass is forced to orbit the Earth faster than its orbital speed would by Newton's laws induce a large tension on the whole tether or cable. The tether expands in size and mass as it approaches the massive counter mass beyond geosynchronous orbit. The tether may start out with a diameter of a few centimeters at the Earth's surface and becomes hundreds of kilometers in diameter at geosynchronous orbit. Along this

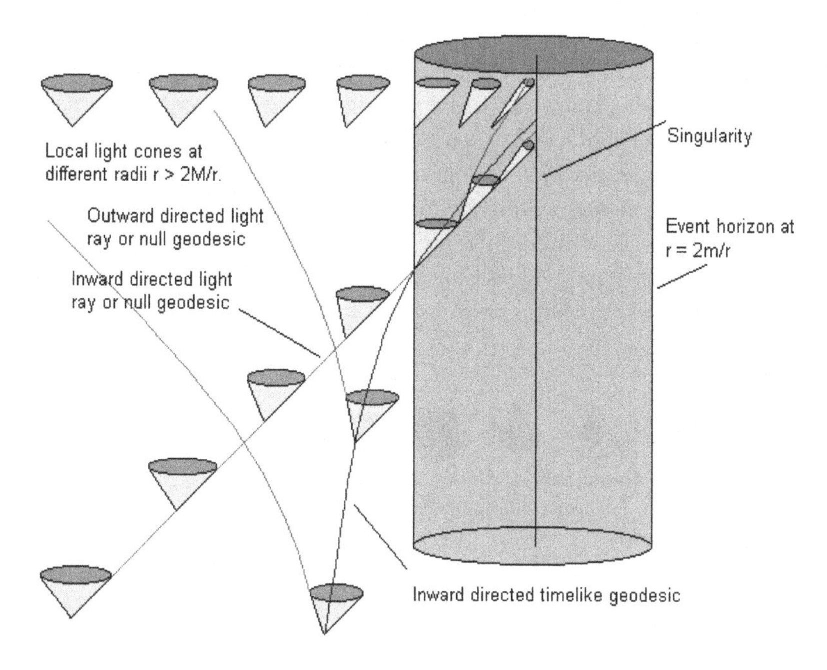

Fig. 9.1. Local light cones in the black hole environment.

tether an elevator could lift people and equipment to geosynchronous orbit. It is proposed that high strength nanotube fibers may be used to weave the tether, but so far nanotubes of this sort have not been developed. All of this requires that millions of tons of material be lofted into space to construct this. Of course this has an obviously very "pie in the sky" sound to it. The scale of this is colossal. Its construction would require the use of many rockets, which the space elevator is meant to replace. So it is hard at this time to seriously contemplate the space elevator as a realistic prospect.

This is the challenge for the starsail. It requires a growing capability in space. The recent history of space flight is not a very good precedent for this. The ISS spacestation and space shuttles are slated for cancellation by 2010. NASA's plans for a return to the moon and missions to Mars faces serious budgetary and political problems. It is at best a 50–50 chance whether this program will take shape. Currently public interest in such space programs is nowhere near what they were in the 1960's during the ramp up to the Apollo lunar missions. However, interest in the unmanned programs remains fairly solid, which so far means some future for space science. It has to be admitted that these have had a far greater scientific payoff than manned spaceflight.

The principal difficulty with the relativistic rocket is antimatter. How does the rocket store antimatter, and where is it acquired to begin with? A lot has to be generated. For a spacecraft that would start out at 200 metric tons and burns up half its mass, this requires 50 metric tons of antimatter. This is an energy equivalent of 4.5×10^{21} joules. This requires as much energy as is generated currently over a 100 year period. This of course ignores thermodynamic losses in generating this antimatter. This is obviously a huge problem that would need to be solved.

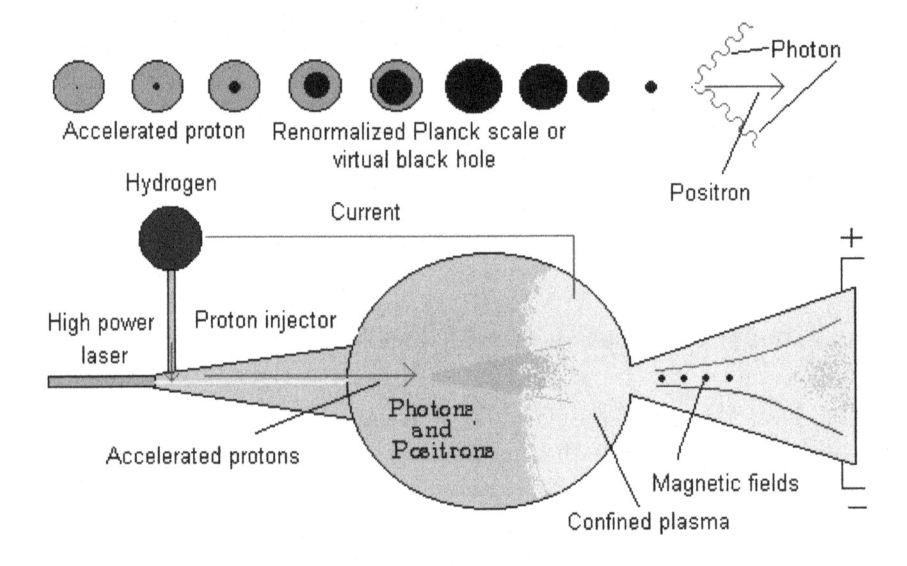

Fig. 9.2. Schematic for renormalizing the Planck length to convert matter to energy.

The problem could be solved only by some exotic physics. In particular if the baryon number can be violated. Such a violation would be $p \rightarrow e^+ + \gamma$. If this could be done then there is no need to generate large amounts of antimatter. The problem is that getting such a violation is not easy. To do this requires study into deeper foundations of physics. Baryons, such as the proton, are composed of quarks. These particles have been confirmed in experiments over the past 30 years. Curiously they are bound in such a way that they can't be released from a proton, or any bound state of quarks [9.2]. Quarks occur in what are called doublets. The first of these doublets is the $\binom{u}{d}$ doublet of the up quark and the down quark. The proton consists of two up quarks and a down quark. The up quark has an electrical charge of $\frac{2}{3}e^+$ and the down quark has a $\frac{1}{3}e^-$ charge. These quarks are bound to each other by the intermediary bosons of the Quantum ChromoDynamic (QCD) gauge field. The analogue of charges for this force are called colors. The QCD force is likely unified with the electromagnetic field and the weak interaction field, responsible for β decay in nuclei, so that at high energy these three forces become embedded into a single gauge field. At high energy a quark and a lepton, which carries a unit charge for the weak interaction, can transform into each other. This is similar to the transformations or symmetries discussed with Newtonian mechanics and relativity. This transformation can convert an up quark and a down quark into a positron. Hence a proton may be converted into pure energy by the process

$$p = \begin{pmatrix} u \\ u \\ d \end{pmatrix} \rightarrow^X e^+ + \bar{\nu}_e + \begin{pmatrix} u \\ \bar{u} \end{pmatrix}, \tag{9.1}$$

where $\bar{u}$ is the anti-up quark and $\bar{\nu}_e$ is an anti-electron-neutrino. The bound state between the up quark and the anti-up quark is a neutral meson which is unstable and decays away, essentially into photons. The X boson exists in the grand unified field theory. The problem is that this process only occurs commonly at energy interactions a trillion times larger than what current high energy machines can reach. Further, this boson will still occur in quantum fluctuations and should give a decay rate for the proton. A proton should have a lifetime of 10^{29} yr, which in principle should be detectable. So far attempts to find the neutrino signature of this decay have been null.

The above describes the $SU(5)$ Grand Unified Theory (GUT), where the designation $SU(5)$ refers to the particular algebraic structure of the above transformations. The failure to detect neutrinos from a proton decay indicates that this particular GUT is probably wrong. However, there are other GUT models. It is likely that on some fundamental level quarks and leptons are inter-changable and their respective gauge theories inter-changable within some unified gauge field. So there must be on a fundamental level a way that the proton can be annihilated so as to violate baryon number. However, this may not be practically done with particle physics. If there is the intermediary X boson, as indicated above, it must be in some fashion produced at energies far lower than 10^{15} GeV, where current particle accelerators are at the 10^3 GeV scale. Quantum field theory is structured around the renormalization group, which is a way of computing parameters on various scales. The renormalization group has its origins with the regularization process in quantum field theory that eliminates infinities. This is a technical issue that we will not explore, for this book is dedicated largely to classical physics. However, it might be the case that subtleties exist with this that might permit physics at a very high energy scale to be rescaled to lower energy by some process. Of course this is pure speculation, and in fact is most likely wrong.

Another way that protons can be destroyed is with black holes. It might sound strange to suggest black holes here, but tiny quantum black holes might be produced in high energy experiments in the near future. If so it is then possible to convert protons directly into energy. A black hole is known to be the collapsed remnants of a star that imploded under its own gravity after its nuclear fuel was exhausted. Such black holes have been found in relative abundance. At the center of galaxies monstrous black holes have also been found. These large astrophysical black holes are obviously outside our ability to produce. Yet tiny black holes in principle can exist as well, where these might be used to generate energy through the direct conversion of matter to energy.

Karl Schwarzschild derived the first solution to Einstein's general relativity in 1916, while serving as an officer in the German army. His solution was for a static (nonrotating) spherical mass, which was later used to derive the perihelion advance in the orbit of Mercury and the lensing effect of the sun. The latter was confirmed in observations of a solar eclipse off Brazil after the end of World War I. His solution contained the factors $\alpha = 1 - 2GM/rc^2$ that enter into the metric line element [9.3]

$$ds^2 = -\alpha dt^2 + \alpha^{-1}dr^2 + r^2(d\theta^2 + \sin^2(\theta)d\phi^2). \tag{9.2}$$

For the sun, the radius is large so that $r \gg 2GM/c^2$ and the curvature near the sun is relatively small. However, in principle this solution exists for $r \leq 2GM/c^2$. For equality this line element "explodes," but this turns out to be a problem with coordinate choice and may be removed by using a different set of coordinates. For $r < 2GM/c^2$ the time and radial parts of the metric change sign. The inward radial direction assumes the role of time. A particle in this region falls inward in much the same way that we progress forward inexorably into the future. Hence escape from inside this region of the black hole is impossible. The line element also diverges to "infinity" as $r \to 0$. This singularity turns out to be real, and not an illusion of coordinate choice.

Schwarzschild died of an illness on the Russian front in 1916, and so never saw his solution employed to confirm a prediction of general relativity.

This solution aroused some controversy, but it was generally thought that the curious properties with the Schwarzschild radius $r = 2GM/c^2$ were a mathematical defect and not physically real. Such mathematical defects occur in physics, and physicists most often ignore them. However, in 1939 Robert Oppenheimer found that this solution did predict that a star which collapses to within the Schwarzschild radius will cease to be observable and its evolution beyond this point is unpredictable. He also found that the singularity at the center is physically real, or at least real within classical physics. This was demonstrated by showing that any material that could resist further collapse in the region $r < 2GM/c^2$ would have to propagate pressure waves at a speed greater than light. As $r \rightarrow 0$ the curvature of spacetime diverges, as does the tidal force on any extended body. Anything that enters the black hole is completely destroyed. As a rule physicists do not like such divergences. Yet here this divergence is surrounded by the Schwarzschild radial limit of observability, called an event horizon, and so this divergence is not observable.

Further mathematical work in general relativity demonstrated that a black hole is characterized by only three physical quantities, its mass, angular momentum and charge. All other physical characteristics of matter and energy that composed the black hole are irretrievably lost. There is a measure of controversy over this, but for our purposes we will say these properties are either destroyed or so completely scrambled as to be erased. This is the "no hair" result [9.3]: no properties of matter that constituted the black hole are observable. Of course if one were to observe something falling towards the black hole it would appear to slow down and any clock on it would be slowed as it progresses towards the event horizon. By corollary light emitted by any object is also redshifted, and become so redshifted as the light source moves towards the horizon so as to make it disappear. So the black hole is a gravity pit that causes things to enter it to leave the observable universe, as seen by any observer falling into the black hole, and to the outside observer things falling into the black hole are redshifted out of view. This also means that a black hole will only acquire mass and not lose mass. Hence the black hole is a self growing eater of matter, destroying anything that enters it.

Black holes have been identified in the universe. Though the black hole is invisible its gravity field exerts effects on its environment. Most often a black hole results from the implosion of a stellar core from a star that explodes off its outer layers in a supernova event. Often these black holes orbit another star, where if this orbit is tight enough material from the star is gravitationally pulled into the black hole. This material becomes ever more heated and energetic as it spirals towards the black hole in an accretion disk. This material emits X-ray radiation that is a signature for the black hole. A number of these have been identified. Mounting evidence indicates that the centers of galaxies contain black holes that have masses equal to several billion solar masses.

There is a bit of a problem with the black hole. A black hole that absorbs a piece of matter with a temperature appears to bury it away, along with the temperature of the mass. This is then a violation of the laws of thermodynamics. A black hole is unable to permit anything from its interior shrouded by the event horizon from escaping. This means that the temperature of a black hole might be thought to be zero. The disappearance of a chunk of mass, with some temperature and entropy, is a violation of the second law of thermodynamics. Further, if the black hole has a zero temperature this is a violation of the third law of thermodynamics. So something is amiss with the black hole. Jacob Bekenstein looked at the black hole as something composed of some indistinguishable set of particles to compute the entropy and temperature of a black hole. His result is rather stunning, for it implies that this gravity sink into oblivion must then emit radiation in order to have a temperature. The resolution of one problem raises another question.

Stephen Hawking demonstrated theoretically how the temperature of a black hole was due to quantum radiance. A black hole emits quanta of radiation spontaneously in much the same way that a nucleus may spontaneously decay. The quanta of radiation emitted by a black hole are photons that quantum mechanically tunnel out of the hole. This tunnelling is an aspect of the Heisenberg uncertainty principle that a spread of energy multiplied by a spread in time, usually the time interval of a measurement, is equal to the unit of action $\hbar = h/2\pi$. Similarly a particle localized in a volume may spontaneously appear elsewhere with some spread of momentum. Thus a tiny unit of mass in a black hole may appear away from the black hole and escape as radiation. This mechanism implies that a black hole will then over time quantum mechanically decay away. The lifetime for a black hole is found to be $T = G^2m^3/\hbar c^4$, which the average solar mass black hole is $T \simeq 6.6 \times 10^{77}$ sec or 2.1×10^{70} years. This of course assumes that the black hole is not absorbing matter. A black hole of a billion grams will decay away in about .1 sec. This continues down to $T \simeq 10^{-43}$ sec for a Planck mass black hole. A Planck mass black hole is one who's Planck wavelength is equal to its Schwarzschild radius. For a $\lambda = mc/\hbar$ and $r = \lambda = 2GM/c^2$

it may be demonstrated that a Planck unit of a black hole has a radius $L_p = \sqrt{G\hbar/c^3} = 1.6 \times 10^{-33}$ cm and a mass 2.2×10^{-5} g. The energy required to probe this region is 10^{19} GeV, which is 16 orders of magnitude greater than what high energy machines are capable of.

The quantum radiance of black holes has resulted in a number of questions central to the nature of quantum mechanics. To understand this on a deeper level various theories within the field of string theory have been advanced to account for the proper statistics for black hole decay. One development which has emerged is that the black hole might emerge at energy much lower than expected. Some theoretical work suggests that so called soft black holes might emerge at energy accessible to the LHC accelerator to be started at Geneva in 2007. I have worked some theory which suggests that the Planck scale might be renormalized to larger scales so that soft black holes might also emerge [9.4][9.5]. Further, this might be associated with the Higgs field. If these or related theories should turn out to be realistic then a proton could in principle be absorbed by this black hole and converted to a positron and other particle-antiparticle pairs or photons.

It is then possible that a matter to energy conversion machine might be developed. This is required to make the relativistic rocket possible. Such a rocket would have to carry such a converter to generate the large amounts of energy required to achieve $\gamma = 12$. Of course this assumes that nature is cooperative with our theories, and even if so that this could be made into a working device. If such technology is developed in the later 21st century or in the 22nd it will be the ultimate source of energy, surpassing anything that nuclear energy could muster. If such a device could be made compact enough then a modest sized relativistic rocket could be constructed to study another star system.

Chapter 10

Electromagnetically Accelerated Nano-bots

A hot topic in advanced technology in these early days of the 21st century is nanotechnology. While the future trajectory of this technology is very uncertain, it is worth writing a short chapter on the prospect for the use of these as interstellar probes. Nanotechnology is the engineering of systems on the molecular or near molecular scale. These systems are computers with an instruction set encoded on molecules, in a way similar to what nature provides with DNA. These instructions are then parsed or translated by other molecules which enter into a chain of reactions. These reactions are envisioned to perform a wide variety of possible tasks, from production of materials to potentially attacking cancer cells. It is not unreasonable to presume that these nano-bots might also turn out to be employed as tiny spacecraft. The advantage with this is that nano-spaceprobes have little mass and so the energy requirements for sending them to low gamma velocities is far less than with a more massive spaceprobe.

Generically these nano-bots would be a form of von Neumann probe [10.1]. These probes would be an extension of nanotechnology, technology on the scale of molecules [10.2]. This field started with the discovery of buckminsterfullerenes [10.3]. The buckminsterfullerene, C_{60}, the third allotrope of carbon, was discovered in 1985 by Robert Curl, Harold Kroto and Richard Smalley. These were found from the laser evaporation of graphite. For this discovery they were awarded the 1996 Nobel Prize in Chemistry. Nano-space probes would require a large enough of a molecular instruction set to be able to take advantage of a range of environments they might encounter at their destination. Upon reaching some asteroid or comet in an extrasolar system they are capable of using the materials available to either reproduce themselves or to fabricate systems and structures on a larger scale. In effect these nano-probes would establish a type of robotic base on

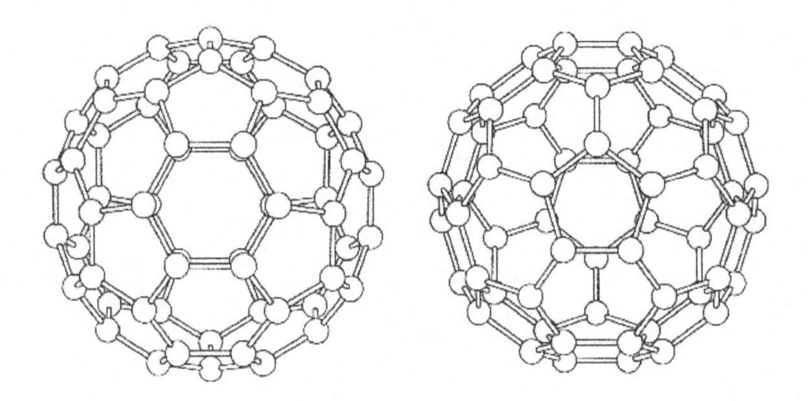

Fig. 10.1. Two views of the C^{60} allotrope of carbon.

an extrasolar asteroid or planetary surface. It is likely that a nano-probe would first reproduce itself to form a type of colony and to be programmed so each nano-bot in the colony is specialized for distinct algorithmic or processing capabilities. This colony then has an emergent complexity far greater than the original nano-probe. From there it might be able to engage in the exploration of the extrasolar system and to send information back to Earth.

Obviously this requires advances in artificial intelligence (AI) far beyond current technological capabilities. It also will require an AI understanding of emergent complexity, self-regulated, self adapting and self-directed growth. Currently this is not well understood, yet it is clear that biological evolution has through selection of genes permitted this to happen with complex living systems. Since this is the case it is not out of bounds to presume that these problems are tractable in some ways and can be reproduced in synthetic molecular computing systems, such as nano-bots.

It is clear that to probe interstellar space with nano-bots it will require that millions of them might have to be sent to an extrasolar system. This is because most of them are likely to miss any sort of target and end up adrift in interstellar space for billions of billions of years. However, maybe a few will find their way to some asteroid or planetary surface where by they can then begin to exploit resources and develop into a complex self-adaptive system capable of exploration.

Clearly these nano-bots are not likely to be rockets. Just as model rockets can't reach space, a nano-rocket is not likely capable of reaching relativistic velocities. So these have to be sent by alternative means.

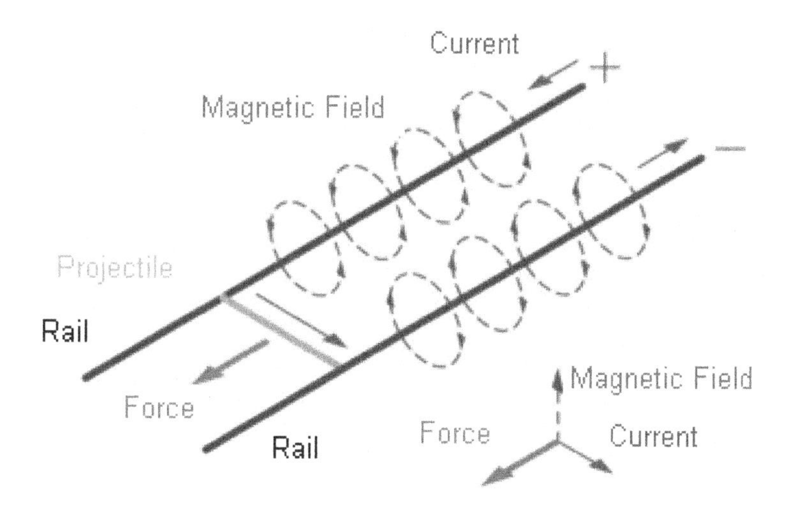

Fig. 10.2. Schematic of a railgun and its electromagnetic physics.

During the Reagan administration the SDI program was initiated, which was an ill-conceived idea that InterContinental Ballistic Missiles (ICBM) could be destroyed before they drop their nuclear payloads on American cities [10.4]. Various ideas were considered for how to shoot down a missile, powerful lasers, particle beams, anti-missile missiles and something called the railgun. The problem with lasers and particle beams is that a target has to absorb the energy from this beam and be thermally deformed to the point of failure. This requires that a laser be pointed continuously on a target spot for some period of time. Conversely a bullet destroys its target not so much be delivering a large amount of energy to it, but by focusing kinetic energy on a spot to cause a material fracture and penetration. A gun does not kill by vaporizing a person, but with a penetrating bullet that passes through a region of tissue, puncturing through vital systems, such as large arteries, the heart or the brain. Further, the rapid shock waves associated with the absorption of a bullet's energy by a target very efficiently delivers that energy in an explosive and lethal way. The railgun was then one of the options for killing a ICBM. This was a way of magnetically accelerating a bullet. Of course for the purpose of space combat this had the deficit that its speed was far slower than a beam of light. SDI turned into a negative sum game, which was then revamped into another organization (BMO) and it continues today. Yet the railgun is a way that nano-bots might be sent to the stars.

A magnetic field induces a force on a moving particle, where this force is perpendicular to the velocity of the particle and perpendicular to the magnetic field. The magnitude of this force is given by the cross product of the velocity of the particle and the magnetic field. This means that if the particle is travelling in a direction parallel to the magnetic field the magnitude of the force is zero, and the magnitude of this force is maximal if the velocity of the particle and the magnetic field are perpendicular to each other. This Lorentz force is then

$$\mathbf{F} = q\mathbf{v} \times \mathbf{B}. \tag{10.1}$$

Of course with Newton's laws the force is $\mathbf{F} = m\mathbf{a}$ from which an acceleration may be computed. This equation involves the motion of a single charge q moving at a velocity $\mathbf{v}$. A current is the time rate of change of a charge $I = dq/dt$ and so this velocity times charge can be replaced with a current. Hence for a current on two wires bridged by a conducting moving object there exists a force on the object given by the Lorentz equation. The diagram 10.2 illustrates the basic set up. This diagram indicates a transverse rod is pushed by the magnetic field behind it induced by the current. Of course the magnetic fields associated with the two conducting wires are in a repulsive situation. For strong enough fields this can cause the railgun to tear itself apart. This limitation did impede these developments for a railgun which could send a bullet to ~ 20 km/sec in order to hit a missile. However, this technique might be used to send a nano-bot to far higher velocities.

A fairly recent development has been with Buckminsterfullerenes, which are C-60, C-120 and higher order carbon bonded convex polyhedra. A topologically different form is that of a carbon nanotube, which has a hexagonal carbon to carbon bond, with p orbitals sticking out from the structure. This is a highly conductive material, which could be configured into wires. These wires might be used to accelerate a nano-bot to very high velocities as a sort of nano-railgun. Due to the dimensions of the problem the force per unit distance on the nano-probe does not have to be that large. Material strengths for carbon nanotubes is considerable, so the gun is not as likely to tear itself apart. The contacts between the nano-probe and the carbon nano-wires requires some addressing, where if the probe is to best to relativistic velocities it can't have actual contact with the nano-wires. The probe must ride above the wires, suspended electromagnetically, and where the current flow between the probe and the wires must be involve electron hopping between the wires and the probe.

An alternative approach would be to put a charge on nano-probes and to accelerate them in much the same way that particle accelerators push protons to relativistic velocities. In this case an electromagnetic pulse in cavities push the charge nano-probe through cavity. There are then quadupole magnets which focus the charge particles into the next cavity. Current accelerators push a proton to $\gamma \sim 1000$. For a hypothetic nano-bot of 1000 daltons a similar system might push it to a $\gamma \sim 2$. For further technical improvements it is not impossible to imagine a nano-bot of some 10^6 daltons being pushed to a $\gamma \sim 2$.

There is of course the obvious problem of what does the nano-probe do when it approaches an extrasolar system. Its large kinetic energy per unit mass must be dissipated somewhere. It might first deploy some sort of molecular thin parachute which breaks it against the photon radiation from the star. However, a quick calculation indicates this is limited. It is possible that the nano-bot could contain a positive charge, which then is repelled by the outgoing charged stellar wind. Of course entering an extrasolar system with its plasma and charge separation of light electrons and protons in the stellar magnetic field is a complex problem. The nano-bot will require the processing capacity to solve this problem and negotiate its entry into the stellar system and reduce its velocity. This is a problem left to the reader and possible researcher who wishes to pursue this problem. Sending a nano-probe to another star is not the principal difficulty, but breaking it so it reaches a destination is rather problematic.

This short chapter was written due to the precedence of Moore's law on the declining cost and expanding ability of processors at a rapid rate. At this time it is very difficult to predict how this type of technology will progress, but there is a prospect that progress forthcoming decades might make this approach the interstellar architecture of choice

Chapter 11

Exotic Propulsion Methods

All afficianados of science fiction are familiar with warp drives, wormholes, transporter beams and other exotic methods. They permit the scripting of a science fiction screenplay without the complexities of relativity and as a way around the light speed barrier. These ideas are not completely without some connection to physics, but the connection is very long and probably thin to the point of being nonexistent. Yet if a warp drive could be developed a spacecraft could travel to another star much more swiftly. It might be possible to travel to another star and return with a week, and if there are faster than light communications (subspace channel of Star Trek), instantaneous communications might be possible as well. Here some exotic forms of propulsion are examined.

The least speculative exotic propulsion system is due to the frame dragging of a black hole. The black hole described in chapter 9 is a static nonrotating black hole. However, since black holes form from the collapse of a stellar core in general black holes are rotating. The result is that points of space near the black hole are pulled around with the rotation [11.1]. In general relativity space is the dynamical field of evolution, where it can morph and change with time. A comparison with the rotation of a fluid near a rotating mass is a fair analogue. All rotating masses will drag space around it. However, for a black hole there is a region near the black hole called the ergosphere, with an outer boundary given by the Schwarzschild radius, where a mass is pulled around inexorably by this effect. This ergosphere is a region outside the event horizon where a rocket with an arbitrary amount of thrust is unable to prevent its being pulled around with the rotation. This is the Kerr solution to the Einstein field equations, due to Roy Kerr. The rocket could in principle escape this region with enough thrust, but it is not able to prevent its being pulled around with the rotation. If this rocket were to then drop a bottom stage towards the black hole the upper part will then acquire a kinetic energy greater than the mass-energy dropped. The spacecraft could then escape the black hole with a highly relativistic velocity. In this case the rocket has absorbed some of the angular momentum of the black hole and energy associated with its rotation.

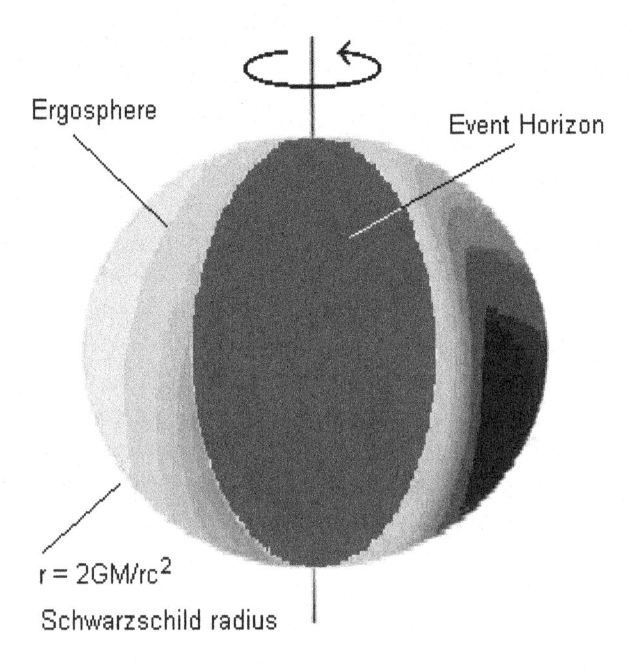

Fig. 11.1. The Kerr solution for the black hole. The contour lines reflect the frame dragging of geodesics in the direction of the rotation.

This "exotic" propulsion system is based on well established physics of black holes. It is certainly an "in principle" propulsion system. However, the closest black hole is hundreds of light years away. So as a practical propulsion system this is not likely, that is unless an unknown black hole is near the solar system or some of the other exotic propulsion systems to be discussed gets the spacecraft to a black hole. Gravity fields also have tidal effects, which in the context of general relativity are the Weyl curvatures. This force along a mass of extension d are seen to be on the order of $F \simeq 2\ \mathrm{GMd/r^3}$. At the event horizon $2GM/r = c^2$. For a stellar mass black hole of 4–15 times the mass of the sun this tidal force would rip apart any mass larger than a meter in extent before reaching the ergosphere.

This means that for any spacecraft larger than a meter a huge black hole is required, such as the black hole at the Milky Way center, where at the Schwarzschild radius $F \sim dc^6/(2\text{GM})^3$ is less. This is thousands of light years away and obviously difficult to reach. Remember, we are mostly focusing on low gamma spacecraft designed to travel $\leq$ $50ly$. Reaching the Milky Way black hole would require a very high gamma rocket for the crew to reach it within a reasonable time. This would have little relevance for the rest of us on Earth, for it would still take many thousands of years on the Earth frame before a spacecraft would reach the galactic center.

So really exotic propulsion requires that the fabric of space or spacetime be deformed in some manner to create a shortcut around the light barrier. The dream is that space might be appropriately distorted so that a spacecraft can bypass most of space or violate the speed of light limit. The first of these is the worm hole. The wormhole is essentially two Schwarzschild solutions glued together at their respective $r = 2GM/c^2$ horizons. This is the Einstein-Rosen bridge that forms a wormhole [11.2]. This forms a bridge or throat between two regions of spacetime. These two regions are connected to each other, so that entering the boundary of one of these regions permits one to exit from an equivalent boundary to some other region. This can happen no matter how remote these two regions are. It is likely that spacetime on a small region near the Planck scale is filled with tiny wormholes that compose a spacetime quantum "foam" . If these exist in the quantum vacuum state it would then outwardly appear possible to excite the vacuum with enough energy to cause one to materialize, even possibly to grow it to a large scale with enough energy. While this might seem to be the case, the issue of quantum gravity may well make this impossible.

The wormhole usable for star travel would be an intra-universe wormhole. Here the two openings exist in the same spacetime. In general wormholes may indeed connect different cosmologies, sometimes called an inter-wormhole. There are similar wormholes that connect D-branes in superstring theory as well. In this case the metric is Euclideanized, where $t^2 \to -t^2$, and the metric is transformed to a standard Euclidean distance in four dimensions. Such a wormhole is called a Euclidean wormhole instead of the Lorentzian wormhole. Figure 11.2 illustrates a wormhole, or Schwarzchild throat. The football space inside is a white hole, forbidden by thermodynamics, or a DeSitter cosmology. In the latter case a virtual wormhole in the vacuum might be the seed for a cosmology. The image on the right illustrates how the wormhole may connect up different regions of spacetime.

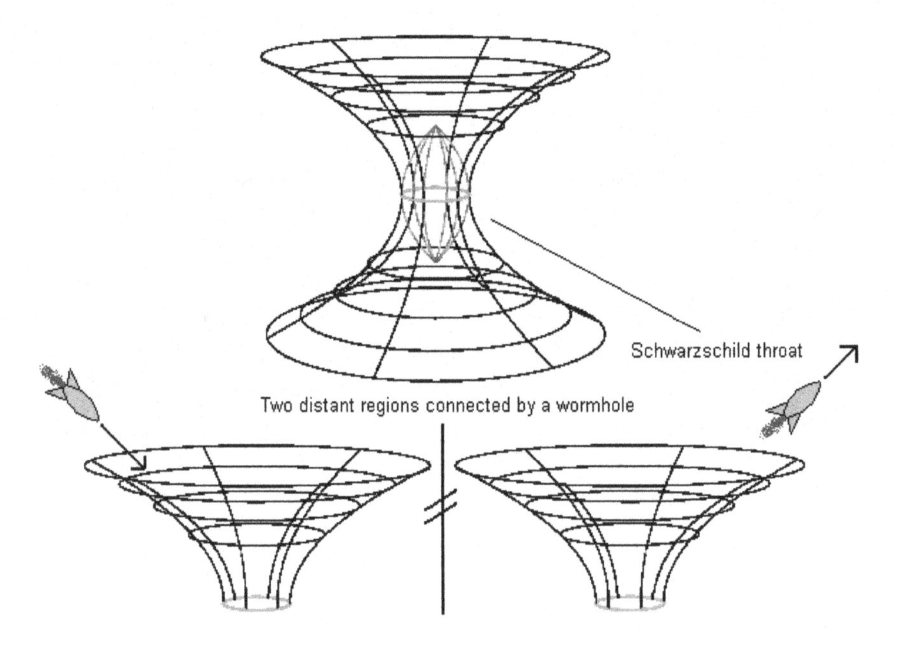

Fig. 11.2. The structure of a wormhole and how it connects distant regions of spacetime.

A strange feature of wormholes is that if one opening of the wormhole is Lorentz boosted to a velocity outward and then brought back the same way it is possible to traverse this wormhole and travel backwards in time. This relies upon the twin paradox problem in special relativity. A twin leaves Earth at a highly relativistic velocity and returns at a high velocity. The total proper time for this twin is less than the coordinate time of the twin who remains on Earth. The same will obtain for clocks near the openings of the wormholes. Once the two clocks have the same time and are separated by a light-like interval, or null line, the region of space in the future of this situation permit closed time-like curves that weave through the wormholes. The light-like separation between the simultaneous clocks defines a Cauchy horizon that separates the time travelling region, called a non-chronal region, from the prior region that is strictly chronal. Hence a traveller who enters a wormhole, returns to the opening she entered and then plunges in again and does this repeatedly, winding through the wormhole endlessly, becomes a time traveller who can only travel backwards in time in the non-chronal region to the future of the Cauchy horizon. The winding loops pile up towards the Cauchy horizon. A traveller who does the same in the past of the Cauchy horizon will also be on a geodesic that piles up towards the Cauchy horizon. This is illustrated in Figure 11.3.

Consider a light pulse that travels through the wormhole openings and then travels to the boosted opening through the region separating them. For light that winds this way through the worm hole in the region past to the Cauchy horizon this light will wind through an infinite number times that pile up near the Cauchy horizon. Similarly light winding through the wormhole in the region future to the Cauchy horizon will pile up towards the Cauchy horizon. This results in a singularity on the Cauchy horizon, where an observer approaching the Cauchy horizon will be hit with a huge pulse of energy that diverges on this horizon.

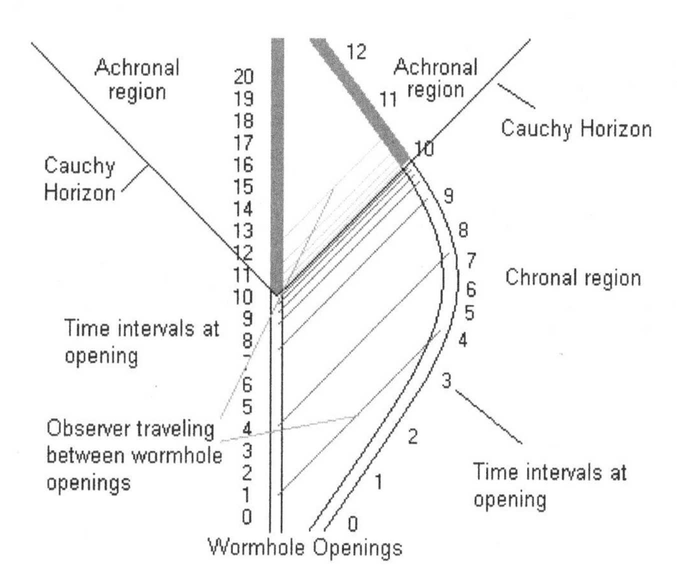

Fig. 11.3. The conversion of a wormhole into a time machine.

This results in a singularity not cloaked by an event horizon. This is cause for some concern. This means as the wormhole is transformed to a time machine that the energy of vacuum modes and fluctuations winding through the wormhole will grow enormously, potentially disrupting the wormhole before the appearance of the Cauchy horizon and the nonchronal region. This naked singularity would undoubtedly have quantum consequences. An unproven statement called the cosmic censorship hypothesis states that such a naked singularity must not exist.

Wormhole solutions also violate something called the averaged weak energy condition. This is a statement that the momentum-energy tensor term T^{00} must be greater than or equal to zero. This negative energy term is required to divert infalling geodesics or particle paths to elsewhere, such as through a wormhole, instead of focusing them further within a black hole. This energy condition, along with the strong and dominant energy conditions, are imposed on solutions to the Einstein field equations to avoid queer solutions, such as the Gödel universe that has recurrent time-like curves. The time machine aspects of the wormhole reflect this violation. Classical gravitation or general relativity permit these violations. Where these violations get into trouble is with quantum mechanics, or quantum field theory. The "stuff" in the momentum-energy tensor are ultimately quantum mechanical, or are quantum fields. If these energy conditions are violated this means that the quantum states of the field are not properly bounded. This results in all types of havoc, where the spacetime will exhibit massive fluctuations. A further problem is the quantum interest conjecture, which states with some measure of reasonableness, that if one attempts to generate a negative T^{00} somewhere that this results in a larger T^{00} elsewhere that overcome the negative T^{00}. Hence if a wormhole could be produced from the vacuum state it may take the equivalent of mass-energy contained in many galaxies. So it appears that this solution might not be realistic. In the end a complete quantum theory of gravity will likely be required to resolve this issue.

Negative energy and $T^{00} < 0$ are permitted in a limited manner with the Casimir effect. The quantum vacuum is filled with virtual quanta, such as photons that pop in and out of a region of space by quantum fluctuations predicted by the Heisenberg uncertainty principle. Two parallel conducting plates will constrain the virtual photons between them to have an integer number of half wavelengths, just as a wave on a guitar string has a specific sets of vibrations and a characteristic note. Because the set of all possible virtual photons outside the plates may have all possible frequencies the vacuum energy between the plates is less. This effect may be measured as a pressure exerted on the plates. So in a limited setting a negative energy, or relative energy, may exist. This appears to give a physical meaning to the vacuum energy, sometimes called the zero point energy. Some people think incorrectly that this is an energy source. The zero point energy is an artifact of quantization procedures, where in fact it might be considered as being ultimately a fiction. Generally this may be removed by certain procedures with no consequence to the physics of interest. Further, the vacuum energy is emerging as something similar to the aether in classical physics. In many ways the vacuum energy is becoming a source of confusion, where it is further known that for accelerated frames the vacuum energy is not separable from particle states. It may well be that a future theory of quantum gravity will change our ideas about the vacuum, where the vacuum energy may be shown to be an illusion.

A wormhole solution has recently been found that does not violate these energy conditions. Yet, just as with any wormhole there is this problem of the singularity on the Cauchy horizon. The wormhole does not have any observational data for its existence, unlike the black hole. A wormhole has certain optical signatures with nearby light that characterize them as distinct from a black hole. Further, if what we think of as a black hole, one observed by its effects on mass-energy falling into it, should turn out to be a wormhole there should be the opposite face with matter and energy spewing from it. So far there have been no convincing arguments for naturally occurring wormholes. Nature usually manages to configure systems that are permitted to exist, such as naturally occurring masers in nebula. Masers are microwave analogues of the laser. So if wormholes can be produced in a laboratory they should exist naturally. While the wormhole has not been ruled out as a natural phenomenon, its status looks to be at best uncertain.

Another science fiction device is the warp drive. Captain Kirk orders, "Warp factor seven."It is true that the fabric of space may in principle be altered to generate a warp drive, at least within a classical setting of general relativity. However, just as with the wormhole this requires a negative T^{00} coefficient, which might render this solution impossible on quantum mechanical grounds. Yet it is interesting to look at this.

Suppose that one had a spaceship with the ability to generate a matter or energy field that could compress a region of space in front of it. This ship is able to shove the points of space in a small region ahead of it into some smaller region. By doing this, just as a carpet is rippled up at some place the rest of the carpet behind this ripple is shoved forward. The ship is then sent along a free falling path towards this bump in the space. But in so doing this moves the bump further ahead. Further, because of this the distance that the ship appears to fall in its frame is shortened by this contraction factor. So if this region is compressed by a factor of 10 the apparent distance the craft falls is a tenth the distance travelled as measured by an external observer. However, to keep this going it is not possible to keep bunching up the rug forever. Slack has to give somewhere. So points behind the ship are stretched out. So this creates a bubble around the ship with points in front that are compressed together and points behind that are expanded apart. So there must be some sort of field that exist around the equatorial regions of this bubble, where the plane of this equator is perpendicular to the direction of motion, that interfaces between the compressed space region and the expanded space region. It is in this region that one requires some type of field with negative T^{00} coefficients. This might doom this as a physical solution and render the warp drive as a permanent denizen of science fiction.

The warp bubble proposed by Miguel Alcubierre is a wave in the structure of spacetime that has the line element

$$ds^2 = -(1 - \beta_i^2)dt^2 + 2\beta_i dx_i dt + \delta_{ij} dx^i dx^j, \tag{11.1}$$

for the index x_i the direction of propagation by the shift vector $\beta_i = -v(t)f(r(t))$ that evolves on spatial surface into the next as the bubble propagates forward [11.3]. The Kronecker symbol δ_{ij} is unity for $i = j$ and zero for $i \neq j$. The radial function $r(t)$ and velocity $v(t)$ are

$$r(t) = \sqrt{(x - x'(t))^2 + y^2 + z^2}, v(t) = \frac{dx'}{dt}, \tag{11.2}$$

where x' is the center of the bubble. The function $f(r(t))$ describes the warping of the spatial surface that generates the warp. To obtain the contraction of space at the front of the bubble and the expansion at the behind this function is

$$f(r') = \frac{\tanh(\sigma(r' + R)) - \tanh(\sigma(r' - R))}{2\tanh(\sigma R)}, \tag{11.3}$$

where R is the radius of the warp bubble and σ is a parameter that might be called the "warp factor". The metric coefficients may be used to compute the curvature coefficients and with the Einstein field equation the momentum-energy coefficients. This reveals that $T^{00} < 0$, which is clearly a potential problem.

If negative energy is permitted and a warp bubble possible, then a warp bubble made large compared to the dimensions of the spacecraft will exhibit curvatures in the central region of the craft which are small. The forward squeezing of space and the leeward expansion means that the region inside the bubble is in a continual free fall. There would be no g-force on a spacecraft in this region. Thus locally the spacecraft is at rest with respect to its local frame. This leads to the curious cancellation of relativistic effects, so there is no Lorentz contraction of the spacecraft length nor is there any change in the observed clock rate. The craft is globally in the same frame as observers on Earth. The compression of space in front of the bubble and the expansion behind have the effect of cancelling out any relativistic effects. An observer on Earth will observe this warp bubble accelerate to velocities beyond the speed of light with no intruding or impeding effects of special relativity. Since the warp bubble is travelling faster than light it cannot be controlled. The walls of the bubble, where the large spacetime curvatures exist, are causally isolated from the interior. As a corollary there is no apparent way that this metric can be generated or engineered to begin with, nor is there any apparent way to turn the bubble off. That this spacecraft is continually falling through what is globally flat space may indicate something of the pathology with this solution. The Alcubierre warp drive has a "something for nothing" element to it, which is removed only if the averaged weak energy condition is restored. So far it appears that the warp drive requires this violation.

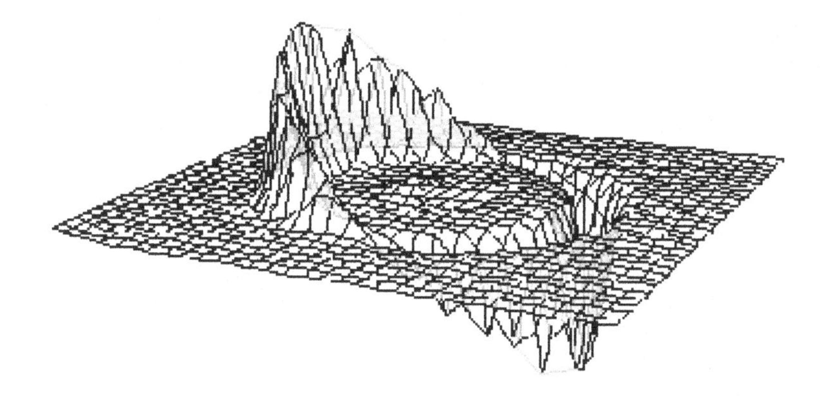

Fig. 11.4. The curvature of space in the Alcubierre warp drive.

Another exotic propulsion proposal is the Krasnikov tube [11.4]. A spacecraft travelling through space in the ordinary manner reconfigures the metric behind it so that time interval along a tube is radically reduced. Thus the craft, while having to travel the ordinary way to its destination is able to rapidly return. This at least can reduce a round trip to half the proper time on board the spacecraft. From the perspective of an observer on Earth the ship returns rather shortly after it left. This curious effect reverses the clock times measured by the Earth observer and the spacecraft observer. A single Krasnikov tube appears to create no pathologies, yet if there are two of them that a craft may loop through time travel, or a closed a timelike loop, is possible. There is again the appearance of a Cauchy horizon with wild quantum fluctuations. These might destroy the Krasnikov tubes. Again this requires that there is some sort of metric engineering capability that is not well understood. It is likely that this approach shares the same theoretical status as the wormhole and warp drive.

There is another problem with these exotic methods of propulsion that would need to be overcome, even if nature permits violations of the averaged weak energy condition. The Einstein field equation tells us that "spacetime curvature stuff" equals "matter-field stuff" by

$$G_{\mu\nu} = \frac{8\pi G}{c^4} T_{\mu\nu}. \tag{11.4}$$

The term $8\pi G/c^4$ is the coupling constant between the physical matter-field and spacetime curvature. This is a very small quantity 2.07×10^{-48} sec^2/g-cm. This is one reason that gravity is a very weak force, where it takes the whole Earth to create the modest acceleration of 10 m/sec^2, and why spacetime curvatures around the Earth and even the sun are quite modest. Thus, in order to engineer a metric one must exploit huge amounts of a matter-field to accomplish this. This coupling constant probably increases in size as the energy of interaction approaches the Planck energy 10^{19} GeV. This is a feature of renormalization in quantum field theory. Yet this requires technology which is able to control a still large amount of mass-energy that is at very high energy. It is clear that this is not technology which will be on the shelf any time in any foreseeable future, if ever.

Quantum mechanics is not a topic of this book, but it does keep coming up here and there. For those who are aware of quantum mechanics and issues involving measurement and entanglement there is something that appears to be a faster than light effect. Two entangled particles that are separated by any distance are correlated instantaneously. For instance two spin particles with a charge in an entangled state, where one particle is subjected to a magnetic field so that it precesses, the other particle, even if very far removed from the magnetic field, will precess also. Similarly if the spin of one particle is measured the second may be instantaneously measured to have the opposite spin. This conjures up ideas that there is some sort of subquantal communication between the two particles. However, John Bell wrote down an equality that would exist if this were the case. Quantum mechanics without any subquantal communication violates this equality. Experimental tests have consistently demonstrated a violation of this equality. It can be said very safely that quantum states in an entanglement separated by any distance do not involve any faster than light communication. So various schemes advanced by people, often not deeply educated in quantum physics, of quantum field effect propulsion systems and the rest represent bogus physics.

Faster than light travel is the holy grail of science fiction and those who envision humanity as a star-faring species in the future. Yet, our current understanding of possible approaches is at best uncertain. There are essentially three approaches for faster than light travel. The first is to ignore special relativity, the second is to abandon causality, and the final choice is to abandon a strong connection between local and global rules of relativity.

The poor man's approach is to ignore special relativity. "It's wrong is all," codifies this approach. The problem here is that special relativity is battle tested. It is so well established that it is not tested so much as it is used. High energy interactions all rely upon the veracity of special relativity. Yet some will persist in their refutation of relativity. Some might cite the problems with reconciling general relativity and quantum mechanics as a reason. Yet this argument is spurious. A relativity refusnick is not dissimilar from a creationist who refuse to accept biological evolution as the proper scientific theory for the relatedness of species. One can only make the case for relativity, or evolution in such related debates, to people in denial in order to convince others listening to the argument. There is a point of no return where such people can't be convinced.

The second approach is to violate causality. This is the case if particles could be converted to tachyons, which are theoretical particles that always travel faster than light. However, tachyons are fields used in superstring theories as something which removes difficulties from the theory by their disappearance. In effect in the early universe they "flew away" to leave the more sane physics behind. Tachyon travel would mean that in one frame a tachyon in motion travels from start to finish, and in another frame it travels from finish to start. Indeed, the Alcubierre warp bubble on the scale of a subatomic particle with some mass carried along with it is essentially a tachyon. Tachyon travel is highly unlikely. In the above case with wormholes and the Krasnikov tubes there is a nonchronal region that can exist under the right conditions. This results in closed timelike curves and time travel. Thus an event that sets in place a series of events may be later effected by this chain that loops back in time. So one could go back in time to kill your self before you started off on a time travel loop. This means physics is illogical. Attempts have been made to show timelike loops do not exhibit these effects, but the arguments are to my sense complicated and "strained."

The wormhole, warp drive and Krasnikov tube separates the local laws of relativity from the global ones. For an observer in the bubble things locally are an inertial frame with all the appropriate properties upheld by relativity. For the distant exterior observer this bubble is a matter-field propagating faster than light. This disconnect between local and global laws of physics is troubling to some physicists. If this were the case it would suggest that we might never be able to deduce much about the universe by using the local laws of physics. Another approach is to say that the speed of light may vary under special circumstances. Spacetime appears to have an absolute cut-off in scale at the Planck length $L_p = 1.6 \times 10^{-33}$ cm, but where this is the same in all reference frames. João Magueijo, Lee Smolin and Giovanni Amelino-Camelia have advanced a modification of special relativity, where near this scale in size, or equivalently near the Planck energy $E_p = 10^{19}$ GeV, the speed of light is variable [11.5]. This theory means that physics of spacetime on a very small scale exhibits departures from spacetime physics on a larger scale. This theory is possible and still considered to be tentatively viable. Yet it is tough to know how this could be applied to get a spacecraft to velocities greater than the speed of light.

This survey of exotic propulsion concepts is meant to illustrate the difficulty in their possibility. As seen above it will be a formidable task to construct low gamma spacecraft to travel to nearby stars. Exotic propulsion concepts are vastly more difficult, and further may be impossible by quantum gravity. It is evident that exotic methods of propulsion can't be seriously considered for the foreseeable future. Low gamma spacecraft exploration of the immediate interstellar neighborhood is speculative enough.

Chapter 12

The Interstellar Neighborhood

This chapter discusses possible destinations for low gamma spacecraft. There are a number of G-class stars accessible to low gamma spacecraft. These might have terrestrial planets worth exploration. Most of the nearby stars are M-class red dwarf stars. It is unlikely that biologically active planets exist around these stars. However, since they make up the largest number of stars in the galaxy they are still systems worth a close examination, possibly by a starcraft.

The stars within 10 light years from the sun are listed below

Star name	Distance ly	Type	M_{sol}	Planet
α Centauri 3	4.36	$G2$, $K0$-1, $M5.5$	1.1, .5, .12	0
Barnard's Star	6.0	$M3.8$	.17	0
Wolf 359	7.8	M	$\sim .1$	0
Lalande 21185	8.3	$M2.1$	.46	3
Sirius 2	8.6	$A0$-1, $DA2$-5	2.14, 1.0	0
Luyten 726-8AB	8.7	$M5.6, M6.0$	.10	2?
Ross 154	9.7	$M3.5$	.17	0

(12.1)

Multiple entries for the star type indicate that the star system consists of two or more stars in a gravitationally bound system. The D-class star associated with Sirius is a white dwarf star, which is the collapsed core of a star. Most of these stars are red dwarf stars. The only star in this immediate neighborhood with identified planets is a dwarf star. It is plausible that such stars might have terrestrial planets, maybe life bearing planets. Though it is likely this is overly speculative, for such a planet may well be tidally locked in the same manner the moon is locked to show one face towards the Earth. Thus one face of the planet would be hot while the other frigid. So we may have to explore further out in order to find star systems that might offer a greater prospect for life. However, even within this small interstellar radius if it is possible to explore all of these stars there might be much to be gained. It is likely that dwarf star systems carry secrets of stellar evolution that warrant close examination. So they should not be excluded from any possible future plan of stellar exploration. However, the most exciting discoveries are likely to come from stars more similar to our sun. Figure 12.1 illustrates the mass distribution of stars.

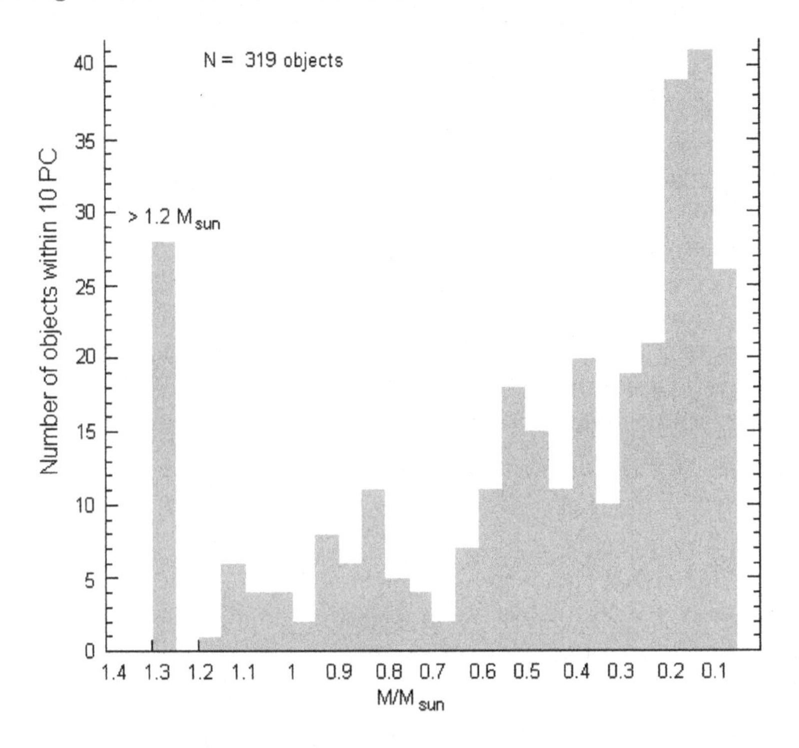

Fig. 12.1. Histogram of stellar masses within 10 parsecs of Earth.

The detection of extrasolar planets had its start with the finding of a gas giant around 51 Pegasus [12.1]. This gas giant is close to the star and its gravitational tugging on this planet more pronounced. Consequently, the star wobbles about the mutual center of gravity for the star and the gas giant. There is a Doppler shift in the spectra of light from the star. Such "torch" gas giants are comparatively easy to find as a result. Gas giants further from the parent star exert a much weaker pull on their star, and the frequency with which the star wobbles is much lower. Jupiter orbits the sun every 11.8 years, which defines the periodicity with which the sun would be observed to Doppler shift according to its wobble. Kepler's law also predicts a much smaller velocity for the sun by $v = \omega r = \sqrt{GM/r}$. Jupiter would be a more difficult gas giant for some extra terrestrial astronomer to find than the 51 Pegasus gas giant.

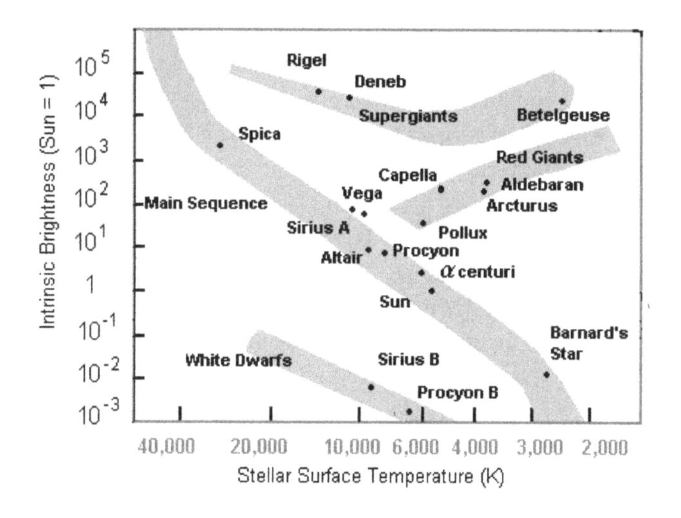

Fig. 12.2. The Hertzsprung-Russell classification chart for stars.

Stars are classified according to something called the Hertzsprung-Russell diagram. This is an empirical chart that relates the luminosity of a star to its surface temperature. From the most luminous to the least the sequence runs as OBAFGKM, with the neumonic "Oh Be A Find Girl Kiss Me." The chart illustrates a definite trend where the higher the luminosity the higher the temperature, where this main band of stars define the main sequence stars. The diagonal lines indicate the diameter of these stars, where an M class star with a large radius is a red giant. There is also a domain for white dwarf stars. It is also clear that on the main sequence there are far more red dwarf stars than there are whiter and hotter stars, such as the sun. It is common to say that the sun is a rather common star, when in fact the most common stars are red dwarf stars. A planet that is biologically active is most likely, at least based upon our local experience, to exist around an F to K class star. The sun is a G-class star which is in what might be the sweet spot for life bearing planets. Higher luminosity stars planet are less apparent, and further these stars have short lives as they consume their nuclear fuel rapidly. So life is very unlikely to exist around any star in the O through A domain. Biological evolution requires lots of time. The situation might be more forgiving for smaller stars that fall into the lower K to upper M range. Of course for such stars with weaker luminosities the orbit of the planet must be much closer to the star. This orbit cannot be so close as to cause the planet to be tidally locked to the star, just as our moon always presents one face to the Earth.

The number of stars that exist within an increasing radius of space within the spiral arms of the galaxy grows very quickly as the cube of the radius. So below is a chart of stars that fit within the F to K range of the Hertzsprung-Russell diagram. For starcraft sent to further destinations these would most likely be the destination of choice. Some of these stars have planets or brown dwarfs that have been identified by their Doppler shift induced by gravitational wobbling. This and other charts below list these stars by name, distance in light years, their mass in solar mass units and how many gas giants or brown dwarfs have been identified. A chart of the nearest stars is:

Name	Distance	Type	Mass	Planet
ϵ Eridani	10.52	$K2V$	.85	1
Procyon	11.40	$F5V\text{-}IV,\ DQZ$	1.5, .6	0
61Cygni AB	11.40	$K3.5Ve,\ K4.7Ve$	.7, .63	0
ϵ Indi	11.82	$K4\text{-}5$	.77	1
τ Ceti	11.88	$G8Vp$	.82	0

. (12.2)

These stars are potentially accessible by a photon sail craft. The star ϵ Eridani is a fairly young star, $< 10^9$ year old, with a planetary disk and one detected planet. So this may reflect a solar system similar to ours in its early phase of development. Procyon is more problematic for containing a biologically active planet. The star has no identified gas giant planets so far. Further, this star has a white dwarf companion in orbit about it, which would be situated around the orbit of Saturn. This would likely be too large a gravitational perturbation for any stable solar system of planets. Procyon is also likely to be too energetic, some 7.5 the luminosity of the sun, for any possible planet with stable chemistry for the emergence of life. ϵ Indi is an orange dwarf star similar to ϵ Eridiani. It further has a brown dwarf star that orbits around it with a mass of .002 the mass of the sun. This is 43 times the mass of Jupiter, and this companion orbits the star at a distance of 1500 astronomical units, or 1500 times the distance from the Earth to the sun. The hunt for planetary bodies closer to this star has been inconclusive. τ Ceti is a star similar to the sun, but is deficient in elements beyond hydrogen and no planets have been discovered. τ Ceti has a dust cloud around it, which suggests asteroids and comets. If τ Ceti has a gas giant it is likely modest in mass and in a large orbit. This star is often the home of ETI in science fiction, but the prospects for a terrestrial planet appears at best uncertain.

Below are F through K stars within a 20 light year distance. Most of them are K-class stars with no appearance of gas giant planets. However, a modest gas giant in a distant orbit could well exist around some of these stars. This list is below

Name	Distance	Type	Mass	Planet
Lacaille 8760	12.9	$K7$	.6	0
Groombridge 1618	15.9	$K7Vne$	.64	0
40 Eridani	16.5	$K1Ve,\ DA4,\ M4.5Ve$	.89, 51, .2	0
70 Ophiuchi	16.6	$K0\text{-}1Ve,\ K5\text{-}6Ve$	.92, .7	0
σ Draconis	18.8	$K0V$	.89	0
η Cassiopeiae	19.4	$G3V,\ K7V$	$.9 - 1.1$, .6	0
J. Herschel 5173	19.7	$K2,\ M4$	.82, .2	0
82 Eridani	19.8	$G5\text{-}8V$	.97	0
δ Pavonis	19.9	$G5\text{-}8V$	1.1	0

(12.3)

A brief description of these stars is given by:

- Lacaille 8760 is a K-class star with no identified gas giant planet. It is also a flare star, with erratic eruptions of energy. This might prevent any bio-active terrestrial planet within the habitable zone of $\sim .16AU$.
- Groombridge 1618 is a K-class star with some similarities to the sun. However, no Jovian planet greater than seven times the mass of Jupiter have been identified.
- 40 Eridani is an orange K-class star with both a white dwarf and a red dwarf companion. So far no planetary bodies have been identified. The white dwarf star is thought to be relatively young. This means that in recent history it blew off its outer layers, which would was a violent event for any possible stellar system of planets. This likely negates any prospect of a biologically active planet.
- 70 Ophiuchi is a double star system of two K-class stars separated by 23 AU. Such a double star may not be conducive for planets, and so far none have been found.
- σ Draconis is a bright K-class star that is somewhat poorer in heavy elements than the sun. So far no gas giant planets have been found.

- η Cassiopeiae is a double star, consisting of a gravitational bounded orbit of a K-class star and a G-class star. These two stars are 23AU in separation, or separated by a distance equal between the orbit of Neptune and Pluto. No Jovian class planets have been identified. The G-class star contains about two thirds the abundance of elements heavier than hydrogen relative to the sun.
- J. Herschel 5173 is a K-class star with a red dwarf companion in a 43AU distant orbit. The K-class star is as abundant in heavier elements as the sun. However, no planets have been detected so far.
- 82 Eridani as a G-class star has almost ideal spectral characteristics for a life bearing planet, However, no gas giant planets have been identified. It also is lower in heavier elements than the sun, which makes it less probable that terrestrial planets exist. It also appears to be drifting off the main sequence and evolving towards its red giant stage of evolution.
- δ Pavonis is almost identical to 82 Eridani. It also appears to be aged and near the end of its main sequence stage of stellar evolution.

This appears to present a paucity of star systems that might bear life. However, none have torched gas giants that could make the orbits of terrestrial planets far more unstable over the duration of the stellar system. Some of these might have modest gas giant planets in large orbits as yet unidentified. If we push further out there are more star candidates to examine. Below is listed only the G-class stars

Name	Distance	Type	Mass	Planet
ξ Boötes	22.1	$G8Ve$, $K4$-$5Ve$	.9–.94, .7	0
μ Cassiopeiae	24.6	$G5VI$	.6–.8	0
ξ Ursae Majoris	27.2	$G0V$, $G5IV$	1.05, .9	4
Chara	27.3	$G0V$	1.1	0
μ Herculis	27.4	$G5IV$	1.1	0
61 *Virginis*	27.8	$G5V$	.96	1
β Comae Berenices	29.9	$G0V$	1.05	0
Groombridge 1830	29.9	$G8VI$	.6	0
κ Ceti	29.9	$G5V$	1.0	0

(12.4)

These stars are described as:

- ξ Boötes is a double star, a G-class and K-class star in a gravitational system. The G-class star is comparatively poor in heavy metals compared to the sun. So far no Jovian type of planet has been detected. The

K-class companion star is rich in heavy elements, which might reflect the environment around the main star. The two stars orbit an average of 33.6AU, which might make stable planetary orbits problematic. However, this is a plausible candidate for terrestrial planets

- μ Cassiopeiae is a small G-class star that is poor in heavier elements. It also has a red dwarf companion. This star is likely to be a halo star as indicated by its velocity. No planets have been found around this star and is a fair to poor candidate for terrestrial planets.
- ξ Ursae Majoris is an interesting double star where both are G-class stars. There are 3 brown dwarf stars identified. This system is also rich in heavier elements. This star system may, unless gravitational perturbations prevent it, be a decent candidate for a search for terrestrial planets. This is a possible candidate for a biologically active planet and a possible destination for an interstellar probe.
- Chara is a metal rich star similar to the sun. While as yet no Jovian planets have been identified around this star it is a decent candidate for a terrestrial planet.
- μ Herculis is an older version of the sun. This star has three red dwarf companions in close orbit, This may make the environment too unstable for there to be planets or terrestrial planets.
- 61 Virginis is one of the best candidates for a search for an terrestrial planet. A 20 times Jovian mass planet has been found, which indicates an active solar system. Further, the star is about as rich in heavy elements as the sun.
- β Comae Berenices is a slightly larger version of the sun. It is also likely to be younger than the sun. So far no gas giants have been found around this star. It is also about twice as abundant in heavy elements. This star may then be a good candidate for a terrestrial planet.
- Groombridge 1830 is a smaller and low luminous G-class star that is very deficient in heavy elements. This makes this star a poor candidate for terrestrial planets.
- κ Ceti is a young G-class star that is rich in heavy elements. It would be a fair candidate for a biologically active planet, one early on its course of evolution, but the star is subject to violent flaring. This likely makes life on any planet impossible.

Very recently a planet about 5 times the mass of Earth with 1.5 times Earth's diameter has been found around the red dwarf star, spectral type M2.5V, Gliese 581 [12.2]. This planet is in the mass range for a terrestrial or rocky planet, and is not likely to be a gas planet. The planet is close enough to this star to have a temperature of 0–40°C, which will support liquid water if the planet has an atmosphere. Because of the mass and dimensions of the planet the gravitational acceleration is 2.2 times that on Earth. This planet is close to the star, .067 AU, and since the star is about 1/3 the mass of the sun this planet orbits this star every 13 Earth days.

Tidal accelerations are $\sim d/r^3$, for d the diameter of the planet and r the radial distance to the large body. The ratio of the tidal force on the moon to that on this planet is

$$\frac{F_{moon}}{F_{pl}} = \frac{m_{moon}}{m_{pl}} \frac{d_{moon}}{d_{pl}} \frac{m_{earth}}{m_{star}} \left(\frac{r_{pl}}{r_{moon}}\right)^3 \tag{12.5}$$

which is about 7.78×10^{-7}, for the subscript *pl* pertaining to the Gliese 581 planet, and the number 1.5 pertains to the ratio of the mass of the planet to the Earth. The radial acceleration that would spin down the planet to a tidal locked rotation would then be about 1.05×10^2 times that which locked the moon. Over billions of years the effect may well slow down the rotation of the planet so it could be tidally locked. If this is so it might make the planet less habitable than presumed, where one face is baked and the other cold. Such a planet might have a belt of life that extends around the limb of the planet as seen from the star, with deserts at the antipodal regions facing towards and away from the star. Life would then exist in a extraterrestrial geographic regions where this star would appear near the horizon.

There is an approximate formula for the time it takes a satellite to become tidally locked. For a satellite of mass m in orbit around a large body with a mass M at a semimajor axis radius a the time it takes for tidal locking is [12.3]

$$T_{lock} \simeq \frac{\omega a^6 I}{3GM^2R^5} \times \frac{Q}{k}. \tag{12.6}$$

Here ω is the initial angular velocity of the satellite. $I \simeq .4mR^2$ the moment of inertia for a spherical satellite, and G the gravitational constant. The constants Q and k are the disspation function and tidal Love number of the satellite. These are generally less certain, but for the moon the ratio of the two is $Q/k \simeq 910$. By inserting numbers for this planet the time for tidal locking is then given by

$$T_{lock} \simeq \omega \times (1.87 \times 10^{17} s^2)\,. \tag{12.7}$$

If this planet is assumed to intially rotate every 24 Earth hours it would then take about $2.16 \times 10^{12}s$, or 6.87×10^4 years, for the planet to spin down into a tidal lock. This clearly indicates that the planet is tidally locked. Certainly now, billions of years after the formation of this stellar system, this planet is tidally locked. This planet was most likely tidally locked within a few hundred thousands of years after its formation. This does not preclude the existence of life, but it would likely mean that life is settled along some zone on the planet where the star Gliese 581 would be seen near the horizon, or just off the horizon.

Since the star is an M class star there is little blue light, so during day the sky would probably be black. I would make the supposition that since this planet has a hot side and a cold side, and if it has an atmosphere, this would mean that these thermal gradients might drive horrendous storms in this zone. The temperature in this goldilocks zone might be comparable to what we have on Earth, but gravity and climate could well make this planet uncomfortable. From the perspective of our comfortable planet Earth this planet is likely a horrid place, at least for us. Even if there is life in this zone, it is likely no more evolved than unicellular forms. It might exist in prokaryotic mats, similar to what the early Earth had as seen with stromatolites. Of course it is difficult know what this planet might be like. A look at early ideas about what Venus might have been like to see that such ideas are too often wrong. Yet if there is water on this planet it might be that a lot of it is in the vapor form due to the hot side. If there is too much water the whole planet might be enshrouded with clouds, and with carbon dioxide this planet might be similar to Venus, but with water vapor in the atmosphere. It could well be a very hot sauna of a planet enshrouded by heat trapping clouds. A close view of the planet by an optical interferometer is the only way coarse grained information about this planet can be obtained. To get a close up view of its geology and the remote prospect for life will require that a robotic probe be deployed on the planet.

The astronomers used the HARP instrument on the European Southern Observatory 3.6 meter telescope in La Sille, Chile to detect this planet. They employed the Doppler shift due to radial velocity of the star, or "wobble," technique. These data are used with Newton's laws to determine the size and mass of a planet, based on small wobble perturbations of the stellar motion off its center by gravitational pull of the planet. The planet is only 20.5 light years away, and so optical interferomters might be able to get a fair look at it. It is further well within the range of a starprobe, and potentially accessible to a photon sail craft.

In the survey of the local stellar neighborhood stars comparable to the sun are mostly featured, while red dwarf stars are largely ignored. This is because M spectral type stars are very numerous, and secondly they are dark horse candidates for biologically active planets. However, they may be of interest in their own right, and their sheer numbers might by the rule of statistics make their planetary systems highly diverse in structure.

The number of G-class stars of course increases the further out one looks. However, this list is cut off at 30 light years, for at best it will take 75 to 100 years from mission launch to the receipt of information from these stars. To reach these 20–30 light year distances and beyond the mission is similar to the Cathedral building in the late middle ages, which took the better part of a century or more to complete. The originators of them never saw their completion. Certainly to probe further out into interstellar space will require such multi-generation efforts. Yet even with the sketchy information that exists there is tantalizing prospects for terrestrial planets. Further, if life is relatively abundant in the universe one of these stars just might have an Earth-like planet. If one is identified by the optical interferometry there will be a scientific value in launching an interstellar probe to examine this planet.

Chapter 13

Will Humans go to the Stars?

It is common for science fiction novels and screenplays to portray starships manned by people, often a rather large number of people. This of course goes back to Jules Verne, where space was seen as something to be explored by people. In the early days of the Soviet and American space programs considerable interest was put forth in piloted spacecraft. The so called race to space was seen as a contest over which side could put astronauts on the moon. It echoed the science fiction screenplays of the 1950s which depicted manned rockets flying to various destinations. Putting humans into space has been seen by many as analogous to the early age of exploration by Europeans. Yet this analogue has its problems.

While the Soviet Union was the first to launch a satellite into orbit and first with a piloted craft in orbit, it was the United States that obtained by first scientific data about space from a spacecraft. Sputnik I went into orbit and sent a repeated beep to indicate to the world its orbital presence. Yuri Gagarin was launched into one orbital cycle of the Earth, but no data about space was obtained. By contrast, Explorer I was launched months after Sputnik and found the Van Allen belts ions trapped in the Earth's magnetic field. So while the popular attention during this time was largely focused on the manned space program, most scientific progress was made in fact by probes.

The success of the Apollo lunar program seemed to suggest that the space frontier was open to humanity. The $L - 5$ society in the wake of the Apollo program devised grand schemes to construct large habitations positioned at Earth-moon Lagrange points. Yet the post-lunar program reality of manned space programs was less stellar. The Soviet Union lofted space stations, culminating in the MIR space station in the 1980s. The major thrust of this program was the study of human physiology in the weightless environment of Earth orbit. The United States in the 1980s began the space shuttle program, which conducted over one hundred missions. However, of these only a few had much impact on space science. The space shuttle program and the subsequent space station program have largely been very expensive and not very productive on scientific results.

There are current initiatives for putting astronauts back on the moon, and to eventually send a crew of astronauts to Mars. As yet these plans are little more than dreams advanced by the President of the United States. Little has left the drawing board to be seriously considered as a working program, and of course any such plan will have to pass the budgetary scrutiny of Congress. It also has to be mentioned that such programs have a consistent history of huge cost overruns. The ISS space station was first advanced as a 10 billion program, and it ended up costing ten times that. Curiously this program is slated for its end within five years. It is likely to have the same ignominious end the MIR station faced, streaking as fire balls through the upper atmosphere. It is very unclear whether the manned missions to the moon will at all yield any scientific results comparably larger than seen with the space shuttle and ISS space station.

A lunar program might serve scientific ends, but the program has to be designed within a direct mission oriented philosophy. It is possible that scientific facilities could be erected and maintained by a program of intermittent astronaut trips to the moon. The programs could involve astronomical instruments based on the moon, such as optical inteferometers and gravity wave interferometers. The moon would provide a solid platform for such facilities. Also the moon has a history connected to the Earth, and lunar exploration may reveal things about the early Earth. However, probably most of this could be done robotically. It is similarly likely that any facilities on the moon could largely be run robotically as well. The moon is also close enough for some of these robots to be run directly from Earth by telepresent means. So within this framework human intervention on the lunar surface would likely be kept to a minimum.

Current plans call for a permanent lunar outpost, with extensive capabilities. This has some unfortunate similarities to the space station. The lunar environment and geology can probably be very well studied with robotic and telepresent capabilities. If the purpose is purely scientific then there is a question why there should be a permanent human presence. Certainly what ever lunar geology astronauts conduct on the moon could be done with robots. Robots do not also have the environmental requirements that people do, such as breathable air, water, food, living space and others.

The space initiative suffers from some issues of mission orientation.

This initiative also calls for astronauts to be landed on Mars. This would be a very expensive program, with current cost projections in the many hundreds of billions of dollars. Obviously humans on the martian surface can perform a far wider range of activities than robots. However, a large number robotic missions designed for specific missions can be done at the cost of a single mission to send astronauts to the Mars. Currently there are two rovers on the martian surface which have conducted a very wide survey. Since they have no metabolic requirements they can also pursue tasks without a return home time limit. Further programs are envisioned for such missions to Mars. A human mission to Mars would be a huge undertaking, which would put a huge investment towards the success of a single mission. Conversely many robotic missions can be conducted for the same cost, but where a low to moderate failure rate can be tolerated. As seen with the Atlantis and Colombia shuttle crashes such catastrophes are enormous, and such a failure with a piloted mission to Mars would be the same multiplied at least ten times.

Some space enthusiasts argue that the future for space flight will mirror the age of exploration by Europeans in the New World. Of course there are a number of striking dissimilarities. The reasons for these ocean voyages had nothing to do initially with any program for exploration or colonization, but to gain access to the markets of Asia. Asia was a source for many materials that Europeans did not produce, such as silk and refined goods beyond the abilities of European artisans. These goods made their way to Europe by the silk road. The silk road was a long caravan route that traversed central Asia and made its way to Constantinople, which later became Istanbul after the Turks took the city. The silk route was prone to attacks by bandits and warring tribes of Mongols and Turkmen. Further, with the rise of the Turkish Ottoman empire these goods faced a large duty imposed on them. The voyage of Columbus was meant to connect Europe with the Asian markets directly and short circuit the silk road. The American continent got in the way of this. The Spanish colonization was instituted as a way to extract gold from Mexico to be used to barter trade in China. These goods were then sold in the European markets. Colonization was a byproduct of this profiteering.

These voyages were conducted because people were already out there. There is nobody out there in space, at least so far as can be seen. The solar system outside of Earth appears to be a complete void when it comes to these sorts of activities. Another feature with the colonization of the

Americas is that the land provided everything needed. In fact there were people already here who were perfectly culturally adapted to life in the Americas. The moon and Mars are the least hostile of any other of the planets to any attempt at colonization, but even still the environments on these planets are horrendously lethal. Literally everything required for life has to be either carried to the planet, or it has to be manufactured there with what resources might exist. Hence a colony on either of these planetary bodies must be a closed self contained system which is able to produce everything needed for the most basic requirements of life. This is possible in principle, but in practice this may not be feasible. In effect for people on a space colony or city the price of everything would reflect the high degree of fabrication required, including the air breathed. In a space city its citizens would have to pay an air bill, which would be rather considerable. The economic capacity of a space city that is able to sustain itself would on a per capita basis be orders of magnitude larger than what currently exists in any city or many nation states. Energy and prepared material do not come cheap, and in space they come at a premium.

Ideas have been advanced for space mining. It states that a space faring economy can be started with the mining of materials. Yet it has to be realized that the price for such raw material commodities are still moderately low here on Earth. Of course once the easy plums have been picked here on Earth prices will start to climb. However, the price for such materials is not likely to ever reach a height to where it will be economical to use highly fabricated materials required for a large spacecraft, plus the energy required for its navigation through space, in order to mine materials in space at a competitive market price. The lunar rocks returned from the Apollo missions have a tiny market value as a raw material source compared to the cost of the Saturn V rocket, and the materials required to construct it. It is likely that raw materials, such as metals that could be mined from space, are going to be more economically obtained from the Earth for a long time in the future. It is likely to be more economical to get metals from the Earth's mantle than from space. The gravity well of Earth is a serious barrier to schemes of easy access to space.

The economy of space has worked only for information. Comsats and other information gathering space systems have worked to some economic advantage. Of course this is a weightless quantity, well nearly so as information is physical. The next weightless quantity that might be had from space is energy. It is maybe possible to put energy collectors in geosynchronous orbit, or elsewhere in the cis-lunar environment, which beam energy back to Earth. As yet such schemes exist only in principle, but they are possible. Might this be the economic source for the colonization of space? Maybe, but it will not be cheap. Electrical energy from space, which requires an expansion of the the electricity grid off Earth, will not come at cheap metered price. Solar photovoltaic energy is still marginally competitive with current energy prices. Expanding this into near Earth space, even there are greater concentrations of solar radiation in space, will not come easily or inexpensively. It is further questionable whether the profit margins, or equivalently the energy surplus, from this will be sufficient to fund or energize a program of human habitation on the moon or elsewhere in the solar system.

So it appears that human expansion into the solar system is problematic. Yet it is still possible in the distant future that humans may step on some planets for a brief period of time. If things continue to progress in a responsible manner it is possible the people will at some time step on the surface of Mars. Of course this assumes that we will not contaminate Earth with any possible martian life form, and visa versa. In order to insure against martian contamination far more robotic missions to Mars are required. Probes capable of returning samples to Earth will be required to put martian material to a rigorous test for biological activity. It is only likely that humans will walk on the surface of Mars only after some of the problems and issues that confront us on Earth are resolved to a measure of sustainability. This puts a "Man on Mars" scenario at best off into the later 21st century.

Putting humans into space is often done with future scenarios of space colonies. A space colony would have to be a mini-Earth capable of supplying all the requirements of its inhabitants. Things that are provided free on Earth, which are being reduced in number as time goes by, must be manufactured. Some schemes involve very large structures with replicas of Earth, complete with forests in some cases. Of course all of this would have to be engineered and maintained in a fail-safe manner. Any failure of a life support system on a space colony could be disastrous. For a large space colony or city the level of complexity would be enormous. Anybody who has maintained a tropical fish tank, or even more a salt water tank, knows that this involves constant intervention and cost. To maintain a terraformed closed system in space would be a vastly expanded version of this.

It is possible that humans will start genetically engineering themselves, which might be required for humanity to colonize space. Again I find this to be a disturbing prospect, yet it has a some possibility or prospect. The economic pressure to have a gene engineered baby with an IQ of 160, or some other attribute, could overwhelm moral principles before long. In such a future children are not born and raised on the basis of unbidden love, but are instead designed commodities. Again something taken for granted becomes a commodity. Indeed those plastic bottles of water so ubiquitous these days illustrates this trend with water. The trends of our current age is that things once considered as "gifts" of nature become later something controlled and commodified. Already preliminary designs for computer-brain implants are being put on the drawing board. It may not be long before parents will have to equip their kids with the latest brain-silicon interface to compete. When it comes to science fiction the movie *GATTACA* takes a fair look at a future of a genetically modified humanity. The only good thing I can think of with respect to myself is that I will likely be dead before this happens. I would prefer that science be considered more as a sort of liberal art, instead of as a system to design increased control methods. This ideal is broken by the fact that a scientific discovery can result in a working device, while a poem does not.

It has to be noted that hybridized and genetically engineered life forms are weaker than their wild type relatives. The degree of care required for genetically modified or hybridized crops and animals is very high. These crops require intensive amounts of fertilizers and in the case of corn the pollen tassels have to be manually picked in order to fertilize the next seed crop. In the midwest August is detasseling season. Take a look at the hybridized cows these days! They hardly even look like cows. Again these hybridizations and now GMOs are done to increase yields (performance), but these organisms are utterly unable to survive on their own as their wild-type relatives. If humanity begins to design its genome much the same may result.

If economic pressures push us in these directions it implies that some subset of humanity may diverge from the rest. The poor are unlikely to have the option for gene engineered babies, and there are a lot more poor in this world than the wealthy. This divergence might push some of future humanity into space in the distant future. However, at this point these humans may be so different as to be not human in our usual sense. They may well be biologically engineered and integrated into bio-engineered and nanotech systems. In effect these neo-humans would be integrated into a control structure of vast proportions and their activities dictated according to its requirements. Again I am glad I will be dead before this happens.

On the other hand humanity may never realize any of this. The coming energy and resource short falls combined with a possible global eco-spasm due to matters such as global warming could put the kabbosh on much of this, which could also cancel any prospect for interstellar probes as well. Even if we manage to avoid this sort of collapse, we may chose to adopt a different mode of life and thinking, one where we chose not to go down this road. In other words we may find that either there are limits on control structures, or we may find pursuing more control something dystopian and to be avoided.

This does not mean that space colonies are impossible, but they are problematic. In effect we already live on a spaceship, one that evolved naturally, and where we evolved to live within it. In many ways it is a spaceship, for it orbits a G-class star, which orbits in a galaxy, a galaxy within a cluster of galaxies, where these clusters are racing away from each other in an expanding universe. The life support system on Earth is not one we need to manage, though recently in order to extract more from it we have increasingly been managing it. However, we largely breathe air and drink water without too much concern. On a space colony this would not be the case. Everything on a space colony would have to be carefully monitored and maintained with the highest of quality control. Failure to do so would run a risk of a complete disaster. Here on Earth, at least so far, we have none of these immediate concern. The question is whether the economic generator of a space colony is capable of supporting that sort of infrastructure. Further, what happens if the space colony either does not do so, or stops being an economic generator and is no longer self sufficient?

The Earth thought of as a spaceship, with a life support system, is likely to garner far more attention than any space colony. In the 21st century it is likely that the health of the planetary life support system will become a major issue. The damage we have done to it will start to effect issues such as agriculture and health. At some point we may find that we will have to restructure our technological world and be forced to engage in regardening ecosystems back to some state of health. Earth will far more likely be the "spaceship life support" issue of the 21st century than the establishment of a lunar city.

In light of this assessment, which is open to disagreement, and there will be those on the pro-space side who will disagree, we examine the prospects that humans will ever travel to the stars. A look at the charts drawn up indicates the time frames involved. The time on board a spaceship that reaches a $\gamma = 2$ are less than Earth time. Further, once the ship has reached a $\gamma = 2$ the time for the non-thrusted flight will be half that on Earth. A crew on board a starship would then spend up to several decades reaching their destination. This of course would be intolerable for most people. Spending the better part of one's life on a small spaceship is not very attractive for most people. It would also likely lead to a social psychology in its crew, similar to what is seen with prisons. It could easily lead to forms of insanity by some of the crew members. Obviously a small or moderate sized spacecraft designed to take people to the stars at a low gamma is not feasible. An interstellar spaceship would have to be huge.

A way around this problem is to get high gammas. A one-gee photon rocket reaches $\gamma = 10$ rather quickly. The chart for accelerations up to one-gee gives

g (m/sec^2)	T	t	d
1.0	28.51	94.76	85.71
2.0	14.25	47.38	42.86
3.0	9.502	31.59	28.57
4.0	7.127	23.69	21.43
5.0	5.701	18.95	17.14
6.0	4.751	15.79	14.29
7.0	4.072	13.53	12.24
8.0	3.563	11.85	10.71
9.0	3.167	10.529	9.523
10.	2.851	9.476	8.571

(13.1)

Thus a hypothetical star 17.42 light years away is reached by the crew on a one-gee rocket in 5.7 years of proper time, if the ship starts to decelerate at one-gee midway in the journey. Of course only 10% of the final craft makes it to $\gamma = 10$ and 10% of that is left to reach the star. The remnant of the spaceship is reminiscent of the small Apollo space capsules that returned for the moon. So this cold equation indicates that the investment would be very large. This tiny remnant of a spacecraft would itself have to be large enough to supply everything required by the crew over this time frame.

A way out of this problem is with the Bussard ramjet [13.1]. A field of radiation ionizes hydrogen in front of the ship, where this field further collects the ions into the opening of the ramjet. Interstellar space consists of 10^{-21} kg/m^3 of hydrogen. In order to collect one gram of ions per second the ramjet scoop must collect them from an area of 10^{18} m^3. If the ship is travelling close to the speed of light, then in one second it travels near 3.0×10^8 m and so the ramjet scoop field must cover a frontal area of about

3.3×10^9 m^2, or 32 kilometers in radius. In a year period this would amount to 5.14×10^5 kg or 315 metric tons. Obviously for a spacecraft which is piloted a far larger amount of ions will have to be swept into the ramjet, since the spacecraft must be proportionately larger. The scoop field will have to be several orders of magnitude larger to provide the mass-energy fuel.

The ramjet was not considered for a spaceprobe. The advantage of reaching a high gamma is of limited value if the intention is to receive a signal back on Earth. If this is the intention then a spacecraft travelling at $.99c$ is of little advantage over one travelling at $.86c$. A high gamma space probe is only of value to a crew on board who take advantage of their contracted proper time. A ramjet with a field capable of sucking in 3.15×10^5 tons per year, which would scoop hydrogen in a cross section of 1800 km, would get a spacecraft of $\simeq 3000$ tons to its destination by accelerating one-gee to a $\gamma = 10$ and decelerating again. A 3000 ton spacecraft could potentially have crew quarters comparable to the ISS space station, with a mass of $\simeq 200$ tons, where the rest of the ship is devoted to the power system and an extensive life support structure. Things might be scaled up from here. It might also be noted that in this approximate analysis nothing has been said about bringing the crew back to Earth. For a 25 year proper time round trip voyage, if they left after graduate school and training, say at the age of 30 they would return at 55+ years of age to a world that has advanced around 40 years. They would return to find their surviving friends about 15 years older than they are. It is pretty clear that the technological requirements for such a Bussard ramjet are extreme, far beyond those illustrated for a relativistic probe.

The other alternative is the interstellar ark [13.2]. This is a ponderous concept. It would be similar to the Orion concept, with some method of propulsion that pushes a huge spacecraft to high velocity, but far slower than the speed of light. The ship would have a mass in the millions to many billions of metric tons. It would house a population of people numbering from several hundred thousands to into the millions. If one were to assume that a single person's requirements were met with a minimum of a thousand tons of equipment and materials very efficiently recycled, a million populated ark would require a mass of at least billion tons. This behemoth ship would make its way through interstellar space at a small fraction of the speed of light. Given that the most interesting G-class stars are at a distance of 30 lightyears or more it would take this ship centuries to reach its destination. It will be the generations dozens of times removed from the initial generation who reach this destination.

This type of ship has appeared in popular fiction. The very campy science fiction movie *Independence Day* has us hapless denizens of Earth visited by something of this sort built by another species of intelligent life who have designs on our planet that don't include us. Of course by wit and fortune we defeat the nasty invaders in the end. The reality of this sort of project is of course completely outside of practical considerations. The mass and energy requirements are on the order of the space elevator, which is something that again is not likely to happen any time in the foreseeable future. It costs a billion dollars or more to send a few ton space craft to Jupiter and Saturn. The economic capacity required to construct such a spacecraft would be expanded proportionately, and even more so since it is designed to travel much faster and to survive as a closed system in the cold of interstellar space for centuries.

Of course the purpose of the interstellar ark is to colonize a planet orbiting another star. In the movie *Independence Day* the aliens came here to live on Earth and to ultimately refurbish their ship or build another one so they could go on to the next biologically active planet. Yet how well does this fiction conform to reality? If we ever identify a biologically active planet around another star, it would be worth sending probe to it. It is presumed we would be precautious and sterilize the probe so as not to contaminate the planet with Earth microbes. In doing so we would accumulate lots of information about the planets's ecology, the molecular structure of its life and other aspects of physiology of organisms. Robots would be deployed on the planetary surface to do the job. Obviously these robots would require a measure of artificial intelligence far beyond current capabilities. On the other hand if you were a space traveller on a landing craft on some life bearing alien surface, would you really open the door and step outside?

Again considering science fiction, the gothic science fiction horror *Alien* films depict a contact between humans and another life form. Here an intelligent, or at least semi-intelligent, life form is found that is completely vicious and wreaks utter havoc. There are obvious flaws in the science, such as the rapid growth of the aliens after they burst out of people, and in one movie a dog, and some other aspects of the alien fictional biology is also questionable, but there is a message here. Such direct contact between humans, and for that matter life on Earth, with some life on another planet could result in horrid consequences. The consequence could easily go both ways. In the *Star Trek* screen plays Kirk, Spock, McCoy and others beam down to a planet wearing their uniforms. They don't even wear spacesuits. In fact there are rather comic bits about inter-species mating, such as Spock is a cross between a Human and a Vulcan. Yet to appear on an alien bioactive planet could easily prove to be disastrous. A human being standing on the surface of a biologically active planet has an immune system evolved for Earth microbes. The biochemistry of alien life could be radically different, indeed it is almost guaranteed to be, and our immune system utterly ill-equipped to deal with such an assault. In effect Kirk and others would likely be like loaves of bread left for mould to gobble up. There might well be extraterrestrial microbes that would "see" humans as lumps of biomass to be consumed. The immune system of the hapless astronauts would respond, but likely not in a way capable of managing exposure to microscopic life forms on an alien planet. Conversely the bacteria and viruses carried by Earthly astronauts might prove to be damaging to life on an alien planet.

Astronauts would only be reasonably sent to an extrasolar terrestrial planet to examine its biology. To colonize such a planet would be a far more daunting task. In the first one risks the prospects the crew will die from some infection their immune system can't manage, with the prospect of biological contamination of the planet. When it comes to colonization it can only be said that the follies of human activity here on Earth could pale in comparison to the difficulties and disasters this would present.

Of course there are those who are chomping at the bit with objections here. And it is agreed that "never say never" has a measure of truth to it. Yet it appears that the scale of engineering required for direct human exploration of the stars, in particular with landings on biologically active planets that might be identified around another star, are very daunting. Such things are not likely to occur. Just as the solar system is probably going to be largely explored by probes and robots, where our contact will be largely by virtual reality, the same will most likely be the case with interstellar exploration.

Chapter 14

Solar System Stability and the Likelihood of Earth-like Planets

Stellar evolution predicts that the Sun will increase its radiation output over a time frame of hundreds of millions to billions of years. This poses an academic question concerning the long term fate of the Earth. If the Earth remains at its current orbital radius over the next one billion years then by the end of that time temperatures on the Earth's surface will become intolerable for the continuation of life. It has been suggested that "planetary engineering" might be able to prevent this by shoving the Earth into an orbit with a larger radius. However, it is argued here that this is not required as the three body interaction of the Sun, Jupiter and Earth will over time cause an outward drift in the Earth's orbit. Further, this drift exhibits $1/f$ behavior and is an indication of a degree of chaotic dynamics in the solar system. This drift on average compensates for the increased heating of the Sun. It is possible that this process has occurred through the existence of the solar system. This remarkable property of our solar system is used to examine what is known about extrasolar systems, and the question on the existence or frequency of occurrence of life on extrasolar planets.

This chapter engages in some rather advanced work with chaos theory in classical mechanics. For those unfamiliar with physics at this depth it is advised you skip over the detail and focus on the results

Korycansky, Laughlin, and Adams [14.1] reported that the orbit of the Earth could be modified through planetary encounters with comets whose orbits are directed into the solar system. It is proposed that energy associated with the orbit of a comet could be transferred to the Earth, and that the comet could then pick up this lost energy through a second orbital encounter with either Jupiter or Saturn. The motivation for this analysis is the fact that the radiation output of a main sequence star increases

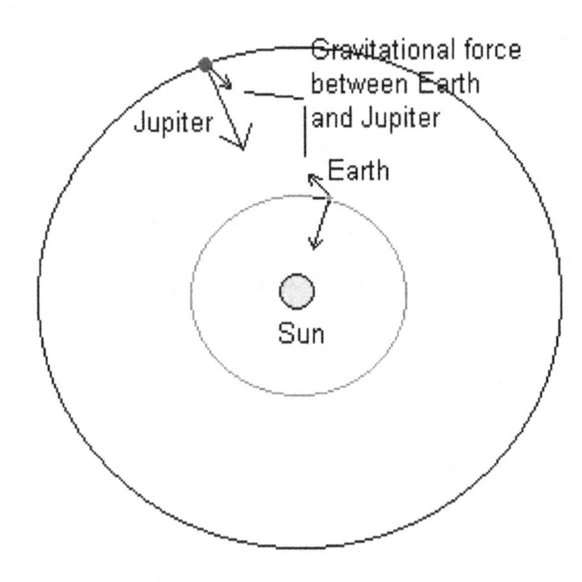

Fig. 14.1. Three body problem for the Sun, Earth and Jupiter.

during its stable existence. In one billion years the 10% increase in the solar radiation will push surface temperatures of the Earth to the boiling point of water, where the Earth's surface environment will shift into one similar seen on the planet Venus. Outside of possible thermophylic prokaryotes (Archeobacteria), the Earth will cease to be the biologically active planet it currently is.

The proposal advanced by Korycansky, Laughlin, and Adams is the Earth's orbit under repeated close encounters with a mass ($\sim 10^{22}g$) on a highly elliptical orbit with a semi-major axis $\sim$ 300 AU every 6000 years may be sufficiently shifted outward over a one billion year period to avoid the over heating of the planetary surface. The 10% increase in radiation output would be compensated by adjusting the orbit of the Earth so that its orbit would exist at 1.05 AU over a billion year period.

From a practical point of view it is highly problematic that such a program will ever be undertaken. The lifespan of more successful civilizations is 500–1000 years, the duration of nation states and empires is often shorter < 500 years, and the duration of the average mammalian species is 3 million years. Further, long term programs are at their longest around a century, as seen with cathedral building during the high middle ages in Europe and monolithic constructions during more ancient periods of history. The modern world engages in much more short term programs than did the ancient world. A long term program on a large scale undertaken in modern history was the 10 year program to put astronauts on the moon, which represents the longer time frames for programs in our age. Thus the long term fate of the Earth most likely does not rely upon humanity to devise such a long term planetary engineering scheme.

However, nature may provide the answer already. During the evolution of proto-life on a molecular level that the Sun would have heated the Earth by approximately 70% the current temperature. This would put the average temperature of the Earth some 3.5 billion years ago at around -60^oC if the Earth were at its current average distance from the Sun. Given that the Earth had a CO_2 atmosphere up until 2.0 billion years ago the early temperature of the Earth would still be below freezing, $\sim -30^oC$ if we assume that the atmospheric pressure was comparable to today's. This has raised debates over the nature of how life evolved out of basic molecules. However, if the Earth were located at .83 AU at that time the temperature of the Earth's surface would be comparable to current average temperatures.

This drift in the average radius of the Earth is a process that is common to 3-body problems. In a 3-body problem it is expected in general that one of the masses will be expelled from the system. This is the so called resonance problem. For a 3-body problem with one huge mass, an intermediate mass and a small mass this process is likely to occur over a long period of time. It is illustrated below that the radial drift in the Earth's orbit and the stochasticity involved are an integral part of a maintaining the Earth at a radius that is appropriate for the continuation of life, presumably for the next 2.5 billion years.

Consider the simplified model of the solar system that includes only the Sun, Earth and Jupiter. The gravitational interaction is the classical Newtonian law of $\mathbf{F} = GMm\mathbf{r}/r^3$. This is a three body problem that in general is not integrable. In this model some simplifications are imposed. The orbits of the Earth and Jupiter are considered to be on the same plane. Further, the small motion of the Sun is ignored. In addition, the initial orbits of Earth and Jupiter are assumed to be circular in order to avoid the ignore perturbation of perihelion advance. In this way perturbations that involve the radial position of the Earth can be exclusively examined. The problem is coded fairly simply and run for $48,000$ Earth orbits. The interaction between Earth and Jupiter perturbs the solar motion of the Earth as a time varying harmonic oscillator force.

It is apparent the Earth wobbles about its Keplerian orbit by a significant percentage of the Earth's radius. On a small time scale this wobble appears comparatively regular. One can consider this wobble as due to a harmonic oscillator in a "first order analysis." The gravitational interaction between Earth and Jupiter is

$$\mathbf{F} = \frac{GM_j m_e}{|r_j - r_e|^3}(\mathbf{r}_j - \mathbf{r}_e). \tag{14.1}$$

First, as an approximation by a binomial expansion, the gravitational force between Jupiter and Earth is,

$$\mathbf{F} \simeq \frac{GM_j m_e}{{r_j}^3}\left(1 + 2\frac{\mathbf{r}_e \cdot \mathbf{r}_j}{r_j^2}\right)(\hat{\mathbf{r}}_j - \hat{\mathbf{r}}_e), \tag{14.2}$$

where $\hat{\mathbf{r}}$ are unit vectors. With $\mathbf{F} = m_e(\ddot{\mathbf{r}}_j - \ddot{\mathbf{r}}_e)$ we find that the motion of the Earth's orbit away from the Keplerian orbit due to the motion of Jupiter is

$$\ddot{r}_e = \frac{GM_j}{{r_j}^3}\left(1 + 2\frac{\mathbf{r}_e \cdot \mathbf{r}_j}{r_j^2}\right)r_e \tag{14.3}$$

which is the differential equation for a harmonic oscillator with a driving force. The frequency of the oscillation from the solution to the unforced differential equation is then seen to be

$$\omega = \sqrt{\frac{2GM_j}{{r_j}^3}} = 3.36 \times 10^{-8}\,\text{sec}^{-1} \tag{14.4}$$

which is a period of 16.7 years. This reflects the periodicity of Jupiter's orbit multiplied by a factor of $\sqrt{2}$. The solution to the inhomogeneous differential equation effectively divides this by $\sqrt{2}$ to give a period of 11.8 years, which is approximately the period of Jupiter's orbit. Since the orbit of Jupiter has a period that has an irrational ratio with the frequency of the Earth's orbit, this Earth's wobble motion irrationally winds around the tori that defines the energy surface in phase space. The solution to this wobbling is illustrated in Figure 14.2.

Now consider the long term evolution of the Earth's orbit. If the motion of the Earth remains on the energy surface the invariant tori is then preserved. This is the essence of the Kolmogorov, Arnold, Moser (KAM) theorem. However, if there is a drift in the motion of the Earth, the tori is then punctured and the motion of the planet can not be described by an equation that is an integrable solution to the equations of motion. The

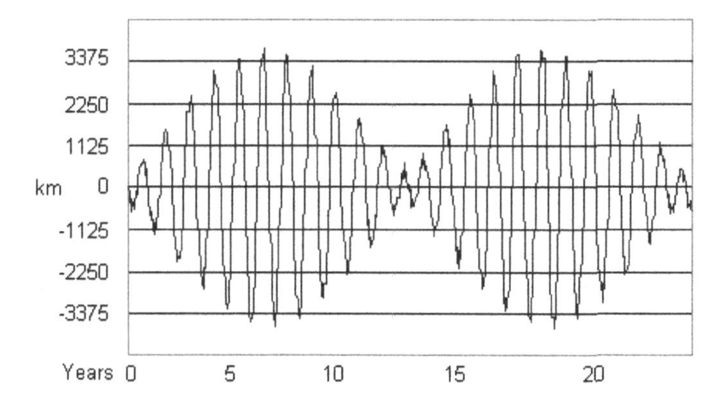

Fig. 14.2. Perturbative wobble of Earth's orbit due to Jupiter.

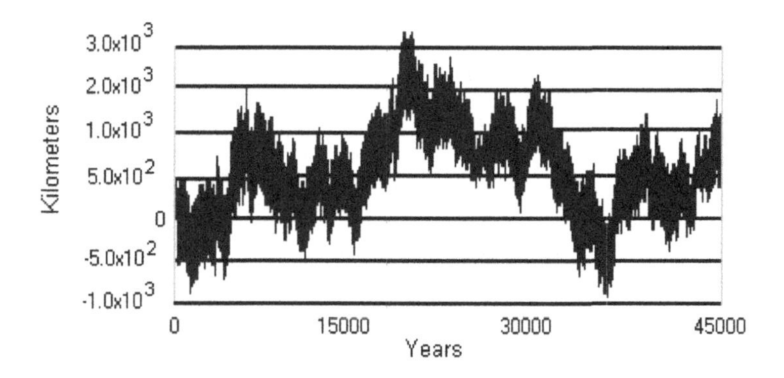

Fig. 14.3. Radial drift of the Earth's orbit over 45,000 years.

numerical result for the long term motion of the Earth is then illustrated on Figure 14.3.

This suggests that over a sufficiently long time period perturbations exhibit puncturing of the KAM surface. With these estimates it is possible to compute the mean drift of the Earth's orbital radius, and further find the average drift outward per year. The drift of the Earth's orbit is then $\sim$ 4.2m away from the Sun annually. Over a billion year period this drift will amount to $\sim 4.20 \times 10^6$ km for the data extrapolated in a linear manner. This would put the Earth at a distance of ~ 1.028 AU in a billion years. This means that for a "constant Sun" the flux of solar radiation would be reduced by $\sim 94.6\%$. If a 10% heating of the Sun is assumed over the next billion years for the Earth at its current orbital radius the increased heating of the Earth's surface will increase by $\sim 4.1\%$. Based upon this the catastrophic heating of the Earth would be pushed back to approximately ~ 2.5 billion years.

The heuristic inclusion of the inward pull of planet Venus the average pull is .21% of the outward pull of Jupiter. This would then reduce the possible biological future of the Earth to 2.05 billion years. One might further then consider the influence of the planet Saturn on the perturbation of the Earth, which might in some manner increase the future biological life of Earth. Of course at this point we are entering an arena of more and more uncertainty. The standard deviation of these data is $\sim \pm 2.5$ m motion annually. This puts the future age of the Earth within the range of 1.15 billion years or until the Sun finally ceases to function as a main sequence star in 5–6 billion years. In the outer range of possibility the solar heating of the Earth will decline by .96% per billion years which, means that the Earth will experience an overall cooling during its 5–6 billion years future.

If there is some overall trend in the climate of the Earth due to its radius from the Sun this should be observable from geological data. The history of climatic conditions on the Earth are relatively well understood during the Cenozoic period. During the earliest Paleocene and Eocene periods the Earth appears to have been comparatively warm, and the apparent absence of large ice sheets in the polar regions [14.2]. During the Oligocene period there appears to be the accumulation of ice on the Antarctic continent. It is most likely that these thermal or climatic fluctuations in Earth history are due to changes in the position of continents from tectonic motion and changes in the albedo of the Earth that had a positive feedback on the climate: the more ice sheets on the Earth the larger the planetary albedo and hence less solar radiation absorbed into the atmospheric and hydrological system of the planet. On a longer time frame it appears that since the Cambrian period that the average temperature of the planet has only varied between 12°–22°C, where currently the Earth's temperature is at the low end of this scale. This is a period of 400 million years [14.3]. If the Earth's radius from the Sun were constant it might be expected that there would be a 4% drift upward in the average temperature of the planet. This is not apparently observed.

It is then reasonable to conjecture that the average orbital radius of the Earth has not been constant in the past. If the average global temperature has been constant over the last 3.5 billion years then the orbital radius of the Earth was .83AU. This implies an average drift during this time period

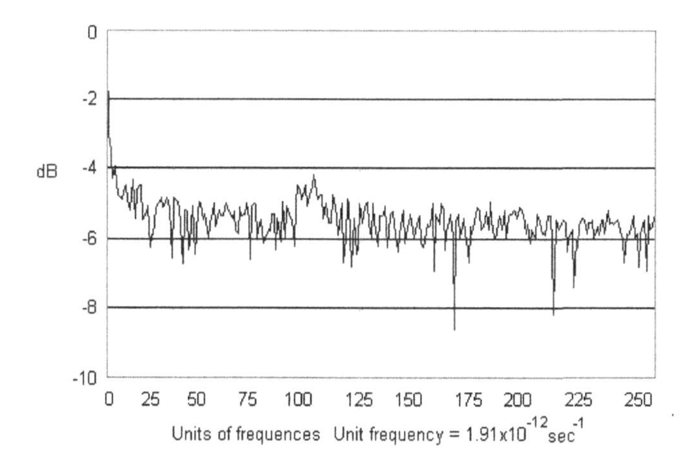

Fig. 14.4. Fourier transform of the radial drift in the Earth's orbit.

of 3.56 meters over each year period as measured currently. This is within these estimates of the current outward drift of the Earth's orbital radius.

This matter leads one into the issue of the stability of the solar system. This was found to be a problem early on in the development of classical mechanics. Newton in his *Principia* wrote on his considerable work on the three body problem of the Sun, Earth and moon. Therein Newton found that the problem of lunar motion appeared to defy capture by an analytic solution for all time. Newton even remarked that God may have to occasionally intervene to keep the entire solar system, which at that time was not known to contain Uranus, Neptune or Pluto (now demoted from the status of a planet), in a stable form. Later the problem of motion stability of planets was examined in 1773 by the mathematical mechanicians Laplace [14.4] and Lagrange [14.5]. This entailed that compound differential equations of motion of planets take into account all disturbances and interference, which amount to around 20000. This also required that resonances in orbital periods and cyclic inequalities in orbits be examined to determine whether the system is unstable or not. The effort by Laplace and Lagrange determined the stability of the solar System within a first order approximation. From a fundamental point of view this is an insufficient answer. It must also be mentioned that these efforts in theoretical mechanics were accompanied by the work of astronomers, opticians and instrument makers that permitted ever refined observations and records on planetary motion.

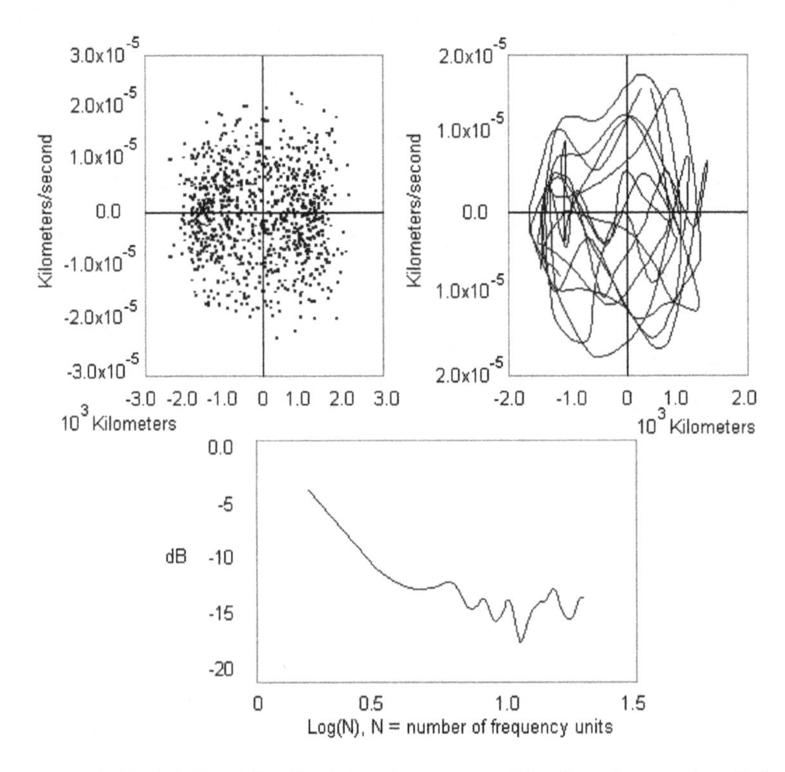

Fig. 14.5. Orbital drift of the Earth in phase space. The first diagram is a Poincaré section and the second is the flow of these points. The third diagram is the log of drift frequencies.

Henri Poincaré [14.6] and Lyapunov [14.7] took up the challenge with the derivation of a mathematically stringent and consistent theory of stability of motion. These developments involve the theory of irregular separatrices in phase space, Poincarë sections in phase space, and the introduction of Lyapunov's exponent as a measure of how two orbits that are initially arbitrarily close in phase space will exponentially diverge in a finite time. Poincarë won the Sweden prize offered for demonstrating solar system stability by illustrating how such stability can not be proven. In the 20th century the mathematician and mechanician V. I. Arnold [14.8] worked to solve the problem of solar system stability. Arnold worked to illustrate that regular orbits of planets in a solar system satisfy the conditions of the KAM theorem. Here with irrational ratios of orbital periods orbits in phase space exist on a torus with irrational winding, that are called the KAM surface.

The irrational winding on a torus is defined by the limit of a continued fraction expansion of various frequencies on regular tori that approach the irrational winding. The KAM surface in a stable theory represents the absolute limit of various tori with rational windings and are associated with integrable solutions. The existence of the KAM surface is seen as the stability of a system with incommensurate periodicities. However, due to overlapping of KAM surfaces or puncturing these integrability conditions can be seen to be less than universal. The breakdown of a KAM surface can lead to motion that is not dynamically predictable, or chaotic dynamics.

There are indications of chaos in the radial drift in the orbital radius of the Earth. First consider a Fourier transform of the time data for this drift. These datum are easily run through a fast Fourier transform. It is then apparent from the data that for frequencies $\leq 3.5 \times 10^{-11}$ sec^{-1} a $1/f$ type of behavior occurs. This indicates on a time scale of $\sim$ 9000 years there is a noisy processes that are exhibited in the dynamics of the Earth's orbit. This means that over this time period and longer there is a net drift in the orbit with a change in the radius in the range $\in$ (15 km, 60 km). This $1/f$ behavior indicates that this drift is an irreversible process, or one that erases some of the information concerning the Earth's orbit every 9000 years. This lead to the conjecture this indicates that the solar system is over a sufficiently long time period not stable, and that the Earth will continue to drift outward until it might be ejected from the solar system. The sun will enter its red giant phase in 5–6 billion years. It is possible that the Earth orbit will have a radius of 1.14 AU during the Sun's red giant phase. So it is likely that the Earth will remain within the solar system while the Sun is a main sequence star. Yet if the Earth survives the red giant phase, it will likely end up in a complex orbital relationship with Jupiter, or be ejected from the solar system altogether within a $\sim 10^{11}$ year time period.

This $1/f$ result is one indication that there is a stochastic process involved with the drift of the Earth's radius from the Sun. Such $1/f$ behavior is associated with noisy processes in electronics and is also existent in other systems. Another signature of chaos is the behavior of a trajectory as it intersects a Poincaré section. This was included in the program, with the plane defined by $y = 0$ and $p_y = 0$ as the coordinate for the intersection of the Kepler orbit. The two graphs below illustrate the Poincarë section for 1000 intersections of the orbit with the Poincaré plane, and the second illustrates the first 100 intersections with the lines connected. This plot involves position and velocity which deviate from Kepler's law. It is apparent that there is utterly no apparent structure to the intersection of this orbit with the Poincaré section. This gives further indication that the drift in the radius of the Earth is chaotic. It appears that for the initial conditions chosen that the dynamics of the drift in the Earth's orbit is determined by a punctured KAM surface.

The logarithm of the low frequency end of the behavior indicates a constant slope in a log-log plot. It is known that $1/f$ noise can be produced through intermittency. Intermittency is where a process that is regular or laminar will experience short periods of irregular behavior. For a scaling parameter small enough the system will exhibit laminar behavior. Then if this parameter is scaled to a sufficiently large value the intermittency is one possible route towards a general mechanism behind $1/f$ noise. Intermittency was investigated as one route to the understanding of stochastic behavior with the Lorentz equations [14.9]. Intermittency is further a property of iterated maps such as the logistics map.

It appears the orbit of the Earth drifts outward from the Sun significantly over a time period of $\sim 10^7$ years. Further it appears that this radial drift exhibits stochastic behavior over a large time scale, and where over a larger time scale the drift averages 4.2 km a year. This may be attributed to the near resonance condition that exists between the orbit of Earth and Jupiter where their orbital frequencies exhibit the ratio $\omega_j/\omega_E = 11.8$. There is the well known vanishing denominator problem for $H(J,\ \theta) = H_0(J,\ \theta) + \epsilon H_1(J,\ \theta)$ for $J = (J_1,\ J_2)$ $\theta = (\theta_1,\ \theta_2)$. Here the generating function written according to the variable J'

$$S(J',\ \theta) = \theta \cdot J' + i\epsilon \sum_{n_1,n_2} \frac{H_{1,n_1,n_2}}{n_1\omega_1(J') + n_2\omega_2(J')} e^{n_1\omega_1(J')+n_2\omega_2(J')} \tag{14.5}$$

will be divergent for the resonant condition $n_1\omega_1(J') + n_2\omega_2(J') = 0$. This resonance condition has lead many to presume that the solar system can not be stable with resonance conditions. Yet the solar system is replete with near resonance conditions.

The number of resonance conditions that exist on the real line are dense. Within any ϵ neighborhood there will exist a countably infinite number of possible resonance conditions that correspond to rational numbers. As the orbit of a planet drifts it will pass through these resonance conditions and be chaotically perturbed. It is to be expected that for simple rational numbers, such as $1/12$ rather than $1003/12000$ strong resonances occur. For more complex rational numbers it might be expected that the instability will be weaker. In other words if the ratio of frequencies are "sufficiently irrational" so that

$$\left|\frac{\omega_1}{\omega_2} - \frac{m}{s}\right| > \frac{k(\epsilon)}{s^{2.5}}, \lim_{\epsilon\to 0} k(\epsilon) \to 0 \tag{14.6}$$

the orbit is more stable. So an orbit that is removed from a "strong resonance" condition near a simple rational number will be more stable than an orbit that is near an orbit with a simple rational ratio of frequencies.

The orbit of the Earth in phase space is then on a torus, with the action-angle variables J_1, J_2, θ_1, θ_2. The action angle variables θ_1 and J_1 are for the Kepler motion of the Earth, while θ_2 and J_2 are the angle that describes the motion of the Earth due to the perturbing influence of Jupiter and the conjugate action variable for the perturbed motion. The action variable associated with the Keplerian motion of the Earth is then

$$J_1 = \frac{1}{2\pi}\oint p_1 dq_1 = \text{const}\frac{R^2}{2}, \tag{14.7}$$

where R is the radius of the planetary orbit and const $\sim \omega_1$ is a constant. From the above discussion on resonance conditions we then have that the ratio of the frequencies $\omega_1/\omega_2 = \alpha(R)$ defines an iterated map,

$$R' = R,\ \theta' = \theta + 2\pi\alpha(R), \tag{14.8}$$

where now θ_2 is set to θ. Now turn on the perturbing Hamiltonian so that $H = H_0 + \epsilon H_1$. There then exist functions $f(R,\ \theta)$ and $g(R,\ \theta)$. An iterated map on the variables R and θ is defined as

$$R_{i+1} = R_i + \epsilon f(R_i,\ \theta_i)\theta_{i+1} = \theta_i + 2\pi\alpha(R_i) + \epsilon g(R_i,\ \theta_i). \tag{14.9}$$

In the case of the θ variable, the function $g(R_i,\ \theta_i)$ is

$$g(R_i,\ \theta) = \int dt \frac{\partial H_1}{\partial J_2}. \tag{14.10}$$

Similarly, with the function involved in the change in the radius of Earth's orbit, we use

$$\delta J = \frac{\partial H_1}{\partial \theta}, \tag{14.11}$$

where this change in the action variable defines the change in the radius of the planet's orbit

$$\delta J = \oint p dq = \text{const}\frac{\delta R^2}{2}. \tag{14.12}$$

It is apparent from numerical analysis that the drift in the radius of the Earth's orbit is stochastic. Consider this function according to iterated maps which determine the stochastic motion on the Poincaré section. Now break this function for the perturbation in the orbital radius as a regular function plus a stochastic part so that

$$f(R_i,\ \theta_i) = f'(R_i,\ \theta_i) + \xi_i. \tag{14.13}$$

Now impose the requirement that the stochastic part of this function ξ_i is a random function that obeys a Gaussian distribution function as,

$$P(\xi_j) = \frac{1}{\sqrt{2\pi\sigma}} e^{-\xi_j^2/2\sigma^2}. \tag{14.14}$$

This distribution of probabilities assumes that fluctuations in the motion of a planet at one time is independent of any fluctuations at earlier or successive time periods. This means the system is Markovian. An integration over all $\{\xi_i\}$ can determine the expectation of the radius R_n

$$\langle R_n \rangle = \frac{1}{\sqrt{2\pi\sigma}} \int \prod_{j=1}^{n} d\xi_j P(\xi_j) R_n. \tag{14.15}$$

Now express the stochastic variable according to $\xi_j = R_j - F(R_{j-1},\ \theta_{j-1})$, which indicates that the stochastic variable is dependent upon the iterated map. The expectation for the radius is

$$\langle R_n \rangle = \frac{1}{\sqrt{2\pi\sigma}} \int \prod_{j=1}^{n} dR_j R_n \exp\big(-(\xi_j)^2/2\sigma^2\big). \tag{14.16}$$

This expectation has the form of a partition function or Euclideanized path integral. Here the dynamics given by $f(F_i,\ \theta_i)$ enters into the path integral, and the stochastic term acts as a diffusive term in this path integral. This indicates the function is explicitly evaluated using renormalization group techniques. This partition function is analogous to that in the Ising model [14.10], but here instead of a set of spins that exist in space there are stochastic kicks that exist in time. These stochastic kicks are assumed to be on average the same, which pertains for the orbit outside of a strong resonance condition.

This path integral can be demonstrated to be similar to the Ising model. For the variation in the stochastic variable $\delta\xi_j = \xi_j - \xi_{j-1}$ the product of any two variations vanish $\delta\xi_i\delta\xi_j \simeq 0$, so that

$$\xi_{i-1}\xi_{j-1} + \xi_i\xi_j = 2\xi_{i-1}\xi_j\,, \tag{14.17}$$

for $i = j$ the sum of these stochastic variables is

$$\frac{1}{2}\sum_{j=0}^{n-1} \xi_j \xi_j = \sum_{j=1}^{(n-1)/2} \xi_{j-1}\xi_j - \frac{1}{2}(\xi_0^2 + \xi_n^2)\,. \tag{14.18}$$

This means there exist additional "endpoint terms" which do not conform to the Ising type of construction. However, for a large enough n this error should be minimal. The expectation is approximately

$$\langle R_n \rangle \simeq \frac{1}{\sqrt{2\pi\sigma}} \int \prod_{j=1}^{(n-1)/2} dR_j R_n \exp\big(-\xi_{j-1}\xi_j\beta\big)\,, \tag{14.19}$$

for $\beta = 1/s\sigma^2$. β is analogous to the Boltzmann factor, but is here a constant is fixed by time the iterated map is run on a computer. As the partition function is invariant under rescaling of β, an approximate renormalization of the coupling constant is

$$\beta' = \tanh^{-1}[(\tanh\beta)^2]\,. \tag{14.20}$$

Here β is a pseudotemperature that gives the time scale for the system. A correlation time or length scale exists that is

$$\tau \propto 1/\log(\tanh\beta)\,. \tag{14.21}$$

For large β this correlation time scale will exhibit a divergence. Physically this indicates that for large β, or long run times, the correlation length becomes very large and that the planet (Earth) may then end up in a region that is exponentially far removed. This suggests that the Earth will after a time either be expelled from the solar system, absorbed into Jupiter, or its orbit may be highly perturbed so that it collides with the future white dwarf remnant of the Sun. It is impossible to predict which of these ultimate fates are in store for the Earth.

As a cautionary note, such renormalization procedures and decimations are not exact. For a repeated series of decimations the error due to the removal of the endpoint terms may cause difficulties if they are large. However, here it is assumed that for a sufficiently large Ising chain these contributions are negligible if the magnitude of each ξ_j is within some reasonable bounds so ξ_0 and ξ_n are not significant contributors to the over all chain.

Physically this suggests that the correlation length is then related to a Lyapunov exponent. The Lyapunov exponent for an iterated map is

$$\lambda(f) = \lim_{n\to\infty} \frac{1}{n} \sum_{i=0}^{n} \log |f'(f_i(R_0)) \,. \tag{14.22}$$

For an iterated map with a coupling strength r the interval $r - r_\infty$, for r_∞ the parameter for instability, the exponent is then

$$\lambda_{f_r} \propto (r - r_\infty)^\gamma \,, \tag{14.23}$$

where $\gamma = \log 2/\log\delta$ is a critical exponent. By adjusting r and δ so that

$$\lambda_{f_r}^{-\gamma} = \text{const} \times (r - r_\infty), \tag{14.24}$$

this defines the boundary between stable and unstable behavior. If the parameter r is identified as the correlation length (time), thinking again of this parameter as analogous to temperature and T_∞ as the critical point, the connection between a Lyapunov exponent and the correlation length is apparent.

Based upon the above Fourier transform and the onset of $1/f$ noise the time for criticality is on the order of 2.09×10^{-11} sec or 6600 years. A criticality for a putative planet at $1AU$ an approximate time interval of drift due to extrasolar gas giant planets exists. It is $t' \simeq (M/M_j)((r_j - R)/(r - R))^2 t$, where M is the mass of a known extrasolar gas giant planet, M_j is the mass of Jupiter, r_j is the orbital radius of Jupiter, $R = 1AU$ is orbital radius of the Earth, and r is the average orbital radius of an extrasolar gas giant planet under consideration. Similarly the parameter β for stochastic behavior is also scaled. This may also be applied to consider the relative influences of Venus and Saturn on the stochastic motion of Earth. This approximate formula gives different correlation times for these planets. The space correlation function for Jupiter doubles at 3 times the rate as a similar function for Venus, and 1.6 times that for Saturn. This leads to an estimate for $1/f$ behavior due to Venus occurs on a scale of $\sim 20,000$ years and for Saturn on a scale of $\sim 10,500$ years. So the influence of these planets on the stochastic motion of the Earth is considerably less.

This discussion indicates that the Earth is in a position that may be relatively unique. The evolution of the solar system may be such that the Earth maintains a fairly constant input of solar radiation during the duration of the Sun as a main sequence star. This likely is important for the evolution of life. The earliest microfossils of life are 3.5 billion years old. The oxygen revolution that started 1.5–2.0 billion years ago and the Cambrian explosion of life 600 million years ago may have occurred under global temperature conditions that are not significantly different than those today. This suggests that the biosphere is not simply an emergent process unique to the planet Earth, but to the configuration and evolution of the entire solar system.

Speculations about life on other planets have been popular since Scarparelli claimed to see *canalli* on the surface of Mars. While life has not been completely ruled out elsewhere in the solar system, it is now almost certain that no planet in the solar system has anything similar to the vast biosphere on Earth. Real information has largely pushed speculation on the existence of life onto extrasolar planets. It is reasonable to argue the star about which another bio-active planet orbits should be similar to the Sun, a G class main sequence star. A more luminous and massive star would exhaust it fuel too quickly, and a smaller star would require the planet to be much closer to the star where its rotation would end up tidally locked with the planetary orbit around the star. Yet main sequence stars evolve by burning hydrogen in a shell that migrates outward from a core of helium, and the star heats up over time. A planet that starts out in the right position for the evolution of unicellular life would then heat up within a $1-2$ billion years and cease to be biologically active without the right gravitational environment established by gas giants. So it poses a question on what types of solar systems might sustain a bio-planet, and what is their frequency of occurrence.

Of course this question is framed in the context of biology as we understand it. This is carbon based biology with the requirements of water in the liquid phase, which should occur under temperatures and pressures comparable to those on the Earth's surface. Speculations of silicon based or alternative chemically based life are not considered here as there exists no information concerning them.

Since 1995 the observation of extrasolar planets has become nearly routine [14.11]. Most of the planets observed are generally rather massive gas giants. So far terrestrial planets are too small to be adequately detected. Yet a survey of these extra solar systems is rather disappointing. Many stars have gas giants in close orbits with periods on the order of 5–20 days to around 1 Earth year. This likely precludes the very existence of terrestrial planets. One possible candidate so far to possess an Earth-like planet with a solar system environment similar to our own is Epsilon Eridani. A gas giant with a mass of .86 times that of Jupiter and an orbital period of 6.85 years has been found. ϵ Eridani is a K2 class star with a mass around .7 that of the Sun [14.12]. This would then put the gas giant at approximately the right position relative to a potential bio-planet situation closed to this star $\sim 5.0 \times 10^7$km. However, tidal locking of such a planet's rotation may prevent any biologically activity on such a planet. Further, the larger perturbation on the orbit of a terrestrial planet due to a gas giant may make the prolonged occurrence of life impossible. So as a candidate for the existence of life ϵ Eridani is probably marginal at best. The recently discovered extra solar system HD 72659 is another potential candidate. The star is classified as at G0 V star with a slightly smaller mass than the sun with 14% less in heavier elements. This star is 167 light years away, which puts it at the extreme range of a possible interstellar probe.

It appears questionable whether there exists an Earth-like planet around any of these stars. It might be best to look for an alternative prospect for the occurrence of a biologically active planet. Ironically a possible answer may lie with the *Star Wars* series of movies. They posit the existence of two biologically active planets that orbit gas giants. Of course gas giants tend to have magnetic fields that trap a considerable amount of high energy charged particle radiation, and a putative terrestrial planet would require a strong magnetic field to shield it. HD 177830 contains a gas giant at an orbital radius comparable to the Earth's orbit. However, this is a K0 star, which means that conditions at this distance from the star are likely to be quite cold. HD 28185 also possesses a gas giant at this radius, and as it is a G5 star temperatures conditions may be right for a possible biologically active moon around this gas giant. However, there is no other gas giant that would perturb the orbit of this identified gas giant to compensate for the increased heating of the star over its lifetime. However, as this orbit is probably comparative stationary this means that the time window for the evolution of life on such a putative moon is narrower than the time that Earth has possessed life.

The stars υ Andromadae and HD 82943 might be a candidate. If a terrestrial planet orbits υ Andromadae b it might be positioned properly for the existence of life. However, as this is an F class star this distance may be outside the boundaries for an a biologically active planet. Temperatures on a terrestrial planetary moon around this gas giant may simply be too high. Yet for a system similar to this one the two other gas giants may act to keep this second gas giant positioned within a radius appropriate for the sustenance of life on a terrestrial planet through the evolution of υ Andromadae. It is doubtful that υ Andromadae b exists in an orbit that is sufficiently stable. HD 82943 has two gas giants b and c, which interact very strongly with each other. Estimates their Lyapunov exponents to that of Earth λ_E for these gas giants around ϵ Andromadae and HD 82943 are $6.3\lambda_E$ and $7.1\lambda_E$ respectively. The measurements of ϵ Eridani are very noisy as this star exhibits variations which make the measurement of this gas giant somewhat problematic. Yet it is possible that a second gas giant exists with a smaller radius around epsilon Eridani with a terrestrial planet. However, the prospects for this type of configuration appears less than likely.

A general analysis of the ratio of the drift times and Lyapunov exponents for putative terrestrial planets around each of the extrasolar systems with that of the Earth is now discussed. The relevant parameter are a ratio of these exponents for each putative terrestrial planet with an orbital radius of $1AU$ with those of the Earth. Current extrasolar planetary results are given in appendix I. The Lyapunov exponents for the orbit of a putative terrestrial planet are in most cases considerably larger than that of the Earth. In many cases the drift rate is 2 orders of magnitude larger than that for the Earth. These analyzes only involve the radial shift of a putative terrestrial planet, and does not include the role of the eccentricity in the orbits of these orbits. In most cases these eccentricities are larger than the orbits of planets in our solar system. This will most likely reduce the prospects for stable orbits of terrestrial planets that could contain life. As such the prospect for a biologically active planet around any of these stars is highly problematic.

A Bayesian estimate for the occurrence of an Earth-like planet may be found at this point. This analysis can only realistically be done to estimate the number of potential orbits comparable to that of the Earth which may exist. Situations, such as with HD 28185 are difficult to study. It is not clear whether a terrestrial planet can form around a gas giant situated ~ 1 AU around the star. Let A be the potential occurrence of a terrestrial planet in an orbit comparable to the orbit of Earth. A indicates that a stable orbit of such a planet occurs. Let B be the occurrence of gas giants that permit the stable orbit of an Earth-like planet. Further, $\bar{A}$ and $\bar{B}$ are the complementary occurrences. Then $P(A|B)$ is the probable occurrence of A given B, $P(A)$ is the probable occurrence of A, and is the Bayesian prior estimate, $P(B|A)$ is the probability of B given A. Bayes theorem gives the equation

$$P(A|B) = P(A)\Big(\frac{P(B|A)}{P(B|A)P(A) + P(B|\bar{A})P(\bar{A})}\Big). \tag{14.25}$$

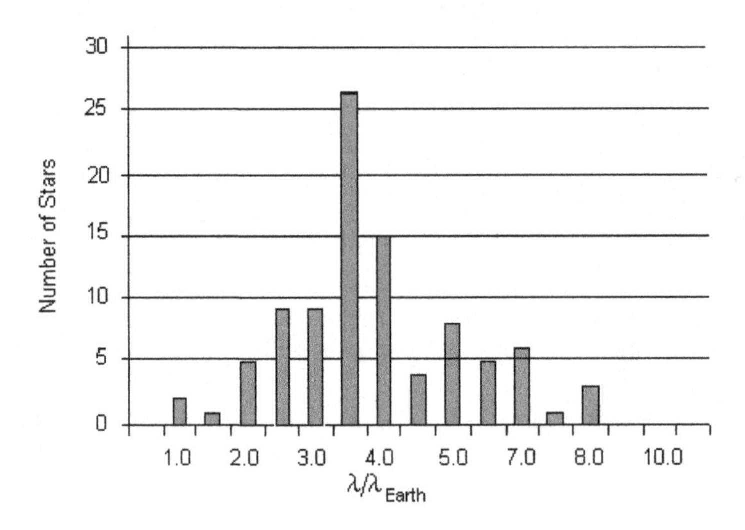

Fig. 14.6. Ratio of Lyapunov exponents between putative 1.0 AU planets and Earth.

To start with, a Bayesian prior estimate is used, which states that out of 145 solar systems examined, only our solar system has an Earth-like planet, $P(A) = \sqrt{1/145} \simeq 0.083$. This is admittedly a Bayesian prior fraught with uncertainty, but is currently the best possible. A much larger data set would be required, and would be bolstered if more extrasolar systems are found which have the proper orbital dynamics of gas giants that can permit the existence of an Earth-like planet. So far outside of our solar system there are only two possible candidates. Yet for the purpose of being conservative these are not included in the Bayseian prior. To find $P(B|A)$ we put the numbers of putative planets with certain range of Lyapunov exponents into bins. A histogram of Lyapunov exponents relative to that of the Earth is presented in Figure 14.6.

To derive a probability distribution for these now sum the number of stars between the Lyapunov exponents in the range $.9\lambda_{Earth} - 1.1\lambda_{Earth}$ this gives a probability of 0.0138. Further we must then factor in the fact that many of these correspond to inward drift. Ignoring the divergent results only 0.27 of these correspond to outward drift. This is then used to obtain $P(B|A) \simeq 0.0037$. Further, since $P(B|A) \ll P(B|\bar{A}) \simeq 1$. we may then approximate the Bayes equation as

$$P(A|B) \simeq P(A)\left(\frac{P(B|A)}{P(\bar{A})}\right), \tag{14.26}$$

and where $P(B|\bar{A}) = .996$. This then gives an estimated probability of $P(A|B) \simeq 3.08 \times 10^{-4}$ for any G-class star containing a bio-active planet. This number is likely to represent an approximate upper bound on the probability for the occurrence of Earth-like planets within extrasolar systems based on current data. The perturbing effects due to the higher eccentricities found in extrasolar gas giants has not been included in this analysis. This further indicates nothing about how solar systems are formed with terrestrial planets. Some extrasolar systems may have gas giants that permit stable orbits for an Earth-like planet, but due to the circumstances of the stellar system's formation no terrestrial planet in fact exists in the proper orbit.

This obviously has an impact on the matter of finding life elsewhere in the universe. It is still possible, though unlikely, that simple life exists in the margins of the environment of Mars. Further suggestions have been made that oceans exist under the ice of various moons around Jupiter. However, while temperatures may be conducive to life, it is unlikely that there is a sufficient flow of energy through the system to permit much in the way of biology. Biology requires a flow through of energy in addition to the appropriate temperature. This means that the search may have to occur outside the solar systems. An optical interferometer may permit the detection of terrestrial planets and chemical signatures of planetary environments may be measured. There is also the SETI program that attempts to measure the radio emissions of extraterrestrial intelligent life forms. This is frankly an ambitious effort. Around .34% of the stars in our galaxy are G-class stars, which means that around 1 in 4.87×10^7 stars in our galaxy comparable to our sun. Further, only 3% of star systems observed have been found to have gas giant planets. This means that approximately 1 in 3.15×10^{-8} stars in our galaxy may have life bearing planets. For our galaxy with approximately 3.0×10^{11} stars, then based upon the analyzes above this means that there may only exist at most around 10,000 extrasolar systems that can permit Earth-like planets in our galaxy. Further, as mentioned above this may represent at best a sort of upper bound, so this number is likely to be reduced from this upper bound. Within a 100ly distance this means there is a 3.0×10^{-4} probability of there being a bio active planet. For a distance slightly larger than 1000 ly the probability of there being a bio-active planet comparable to Earth is approximately unity.

This does not preclude all possible bio-active planets. There may be many terrestrial planets with proto-biological evolution taking place which will never completely mature to any level near what is seen on Earth. This may have happened on Mars. So there may in fact be many more terrestrial planets that exhibit proto-biology or life on a basis similar to the early PreCambrian period on Earth, but are not in stable enough orbits to sustain this over a long period of time.

An article by Gonzalez, Browlee, and Ward [14.13] indicates that there are zones within galaxies where habitable planets may occur. This zone accounts for a band midway between the core and outer reaches of a galaxy, and includes around 10% of the galaxy. This will likely imply that the number of possible terrestrial planets similar to Earth will be around 1000 within the entire Milky Way galaxy. It further should be stressed that these estimates above are upper bounds. During the work on this project new extrasolar system data has had to be included in the analysis twice. These analyzes have been repeated several times as new information on extrasolar systems has materialized. On average the above upper bound has declined. This would indicate that as yet the data is insufficient to have as yet arrived at an upper bound that is constant in the face of new extrasolar system data. It is possible that estimates of these upper bounds will decline for some time before they becomes "stable." This will require reasonable data on extrasolar systems with $\lambda/\lambda_{Earth} \sim 1.0$, which could gravitationally support a putative Earth-like planet in a prolonged stable orbit. Whether or not a terrestrial planet exists around an extrasolar system will obviously involve complexities that are beyond the scope here. This suggests the number of Earth-like planets in our galaxy could be a few hundred or less.

It appears that the prospects are probably not terribly good that the Search for ExtraTerrestrial Intelligence (SETI) program will be detect an ETI signal within our galaxy. Given that these G-class stars will be in various stages in their evolution many of them may be younger than the Sun and others older. Out of $\sim$ 1000 biologically active planets that may exist in the galaxy, some may have early life forms and others may be late in the evolution of life. It is likely that if all of these planets could be found and their chemical signatures analyzed that we would detect these planets in their various stages of evolution. As a rule complex biological organisms, such as mammals, persist for 3 million years on average before extinction. There is also little evidence that evolution operates towards any sort of preferred direction. With the input of Gibbs free energy from the Sun it might be argued that this permits an increased complexity in life forms. This complexity is exhibited in the growth of tissue types, diversity in species variation, and behavioral complexity. However, there does not appear to be any "formula" that demands that biological evolution must give rise to intelligent life forms, where that intelligence is defined by a linguistic ability, with the capacity to abstract concepts and express them to others of the same species. Further, Homo sapiens has existed for approximately $100,000$ years, which makes our tenure on this planet quite recent on a geological time scale, and given various problems we have created for ourselves it can be questioned how long we will continue. This means that our existence on this planet is 0.0002 of the paleontological history of life, and our history where we have had electronics technology is $2. \times 10^{-7}$ of the this history. It might be presumed that an ETI able to develop technologies may face the same issues of resource depletion along with planetary entropy in ecosystem destruction and contamination to one degree or the other that our species faces. So it could be posited that intelligent life is not a terribly persistent life form in the universe. This may mean that the "time window" whereby a SETI program can detect an ETI is very narrow. This narrowness is still quite constrained even if a few ETI can exist for $\sim 1.0 \times 10^6$ years as a technically advance life form.

This obviously leads into the question of interstellar travel. It might be argued that a sufficiently "wise" intelligent life form may exist so they could in time be able to engage in interstellar travel. While this is not impossible, it is terribly problematic. Such programs are easily argued as exceedingly difficult. Interstellar probes are more likely than direct star travel. Beyond that, ideas of spacetime configurations, such as worm holes, warp drives, and the rest are difficult to physically argue for. These types of solutions are admitted by the Einstein formalism of general relativity. However, these solutions require a negative energy condition on the source of the gravity field. This leads to some serious problems with the quantum mechanical basis of a putative negative energy field required to permit this type of spacetime solution. Further, it is the opinion of the author that such solutions will be physically eliminated from a consistent theory of quantum gravity, in just the same manner that quantum theory eliminated some spurious predictions of classical electrodynamics. This tends to suggest that if star travel is performed by an ETI, or possibly by ourselves in the future, it will be done with relativistic rocketry, hibernation, multi-generation starship arcs and other structures that are in compliance with the theory of relativity's demand on the invariance of the speed of light. This suggests that star travel is a terribly difficult process to engage in. As such it is unlikely that various ETIs that might exist are ever to travel to each other's solar systems. This means that combined with the relative scarcity of biologically active planets that exist in the galaxy, with the improbability that ETIs exist close to each other at the same time as defined on the Hubble time frame of the universe, that is is incredibly unlikely that the Earth is being surveyed by spacecraft that originate from some extraterrestrial life form.

Chapter 15

Life on Earth and in the Universe

In this final chapter a number of issues are mentioned, admittedly with a bit of a broad brush. Largely this book was written as a way of presenting some aspects of basic physics in a novel format. It is not certain whether interstellar probes will ever become reality. In order for star probes to become reality it is incumbent that certain conditions exist, including the survival of the human race. It also requires that there be no "end of science," which has been a concern voiced by a few. Interstellar probes might, if they come to fruition, enter into a larger picture of what humanity may learn about the universe and life within it.

The primary purpose for sending a probe to another star is to address the issue of life in the universe. If it should be found by optical inteferometric means that all planets within a 10 to 50 light year range are biologically inactive the interest in sending probes will be muted. Such dead planets would either be too hot, such as a Venusian hot house, or too cold. There might be some planets that fall into a grey zone, such as Mars does today. However, in a future time we may have a better sense about the state of a Mars-like planet. If optically identified terrestrial planets appear dead the dynamics or nature of an extrasolar system it may be well enough examined remotely.

If a terrestrial planet is found to have an oxygen atmosphere with traces of methane this would be a sure sign that the planet is similar to Earth. Optical remote viewing of this planet will never be able to resolve issues about the nature of this life. Something has to reach the ground on that planet and perform experiments. Would these organisms have DNA? Are they organized in ways similar to life on Earth? Or are there completely alternative evolutionary trajectories with macromolecular structures of life? Are there completely different ecological systems that exist on such planets?

There is another possible class of planets that we might find by optical means. This would be a planet that is similar to what is thought to have existed on the early Earth at a point where self organizing molecular systems were just integrating together. Such a planet would exist around a rather young star, have a carbon dioxide atmosphere with organic molecular clouds and have a modest temperature. β Comae Berenices is a possibility for such a planet. A probe sent to such a planet may reveal the conditions under which life began on Earth and quantitative analysis of conditions might indicate something of the chemistry of proto-biology.

Contrary to popular thinking biological evolution does not say anything about the origins of life. Darwin laid down the basic theory on the interrelatedness between species of life. This theory has been consistently supported by preponderance of evidence, and further extended as well to successfully include molecular biology. The biological theory of evolution is as well established scientifically as Newton's laws. However, this does not tell us how the earliest forms of life emerged from chemistry. It appears that evidence for this on Earth has been erased by the activity of life subsequent to its emergence. To find evidence for the nature on how life emerged we may well have to look outside the Earth.

There is a prospect that probes sent to Mars will reveal chemistry of proto-life that might have started, but got frozen out as the planet evolved into a state incommensurate for life. If this is not found on Mars, an obviously closer and more convenient planet to study than an extrasolar planet, then evidence for conditions leading to proto-life might have to be found outside the solar system. There are speculations that life might exist in the great oceans under the ice crusts of Jovian moons, such as Europa. Of course this is something that will have to be tested later in the 21st century. However, there is a problem of energy flow. Even if liquid water exists under ice crusts on these moons the energy flux through the system may simply be too small for there to be much dynamic chemical activity for life.

Of course at this time the uncertainty of things permit all types of speculation. If knowledge about extrasolar systems expands over the coming decades the room for such speculations will narrow. Yet the purpose of such extrasolar probes would be to address one of the most basic questions, the origin and nature of life. This question is as basic as the nature and origin of the universe. It is the call of the *Star Trek* series with its motto, "Seek out new life and new civilizations." Though it is unlikely that civilizations will be found this way.

If we find by optical means a planet that is biologically active, send a probe there and then find there are life forms organized in a technological fashion, even if primitive by our standards, that would of course be the ultimate find. Maybe they would be very technically advanced, but subtle and careful in how they conduct themselves. This may be what we humans will have to do to survive as well. However, such a find is extremely unlikely. Extra Terrestrial Intelligence will only likely be found by radio contact, which could be across hundreds or thousands of light years. The rarity of intelligent life makes it unlikely that we will contact them through a probe we send to another star. The Search for Extra Terrestrial Intelligence, as an attempt to find evidence of intelligent life through radio signals they might be transmitting, so far has produced null results.

Currently we are experiencing a flowering of astronomy and a great leap forward in our understanding of cosmology. This has included a growing understanding of extrasolar systems. This understanding is in its infancy, and techniques with optical interferometry may well reveal far greater details about these extrasolar systems. Additional programs may come forth using ion propulsion and the VASIMR to explore our own solar system to a greater depth as well. Gravity wave interferometers may give a final confirmation of Einstein's general relativity and even serve as telescopes. This gravity telescope may permit us to look at the earliest moments in the big bang cosmology. Will this continue on into the future? It is certainly hoped this will be the case. If this progress continues it is possible that within a century interstellar probes might start to ply their way to other stars.

When I started this book a disastrous hurricane reduced most of New Orleans into a soggy ruin. There are indications that this may have been due to the global warming effect from the emissions of carbon dioxide from fossil fuel use. There are also indications that the world's oil supply may be approaching its half way point, where energy output is expected to decline beyond that point. For our current world so heavily dependent upon petroleum this may cause severe infrastructure and economic problems, which might have a negative impact on continued scientific progress further into the 21st century. If this does turn out to be the case future schemes for probing other stars will be delayed or permanently prevented.

It is my opinion that these problems need not cause a halt to such progress, nor do I think that this will push our world into some dark age. It is emphasized that I am saying "need not cause," but it certainly might cause this. A failure to engage these problems in an appropriate manner may well give these problems their blind license to push us into a state of ruin. This requires a realistic understanding of these problems, which sadly lacking and urgently needed. In order to have this a realistic view of the world is a prerequisite.

Currently a pseudo-scientific idea is being promoted as an equal alternative to biological evolution to be applied in school curricula. This idea, called Intelligent Design, erects the principle of "Irredicible Complexity," as an argument for the existence of some intelligent designer, a code word for a God, as the source of biological diversity. The idea is that complex structures seen in life have some basic level of complexity which can't be derived from simpler structures. However, molecular biology indicates so far how various complex peptides emerge from simpler ones in variations or as dimers, tetramers and so forth. No structure in biology has been found to be completely irreducibly complex. This idea has failed to gain any body of peer reviewed literature behind it. Further, the theory of biological evolution, as initially outlined by Charles Darwin, continues to be upheld by a vast preponderance of observations. In spite of this the Intelligent Design pseudo-theory has a lot of political support, including from the President of the United States. This President and his base of support also deny any connection between human activity and the current warming of the global climate.

This reflects a crisis of thinking and world views in American society, which so far does not seem to afflict other developed nations to this extent. In spite of the overwhelming support for biological evolution about half of Americans believe in Genesis as literally written. The big bang model, supported by a growing weight of evidence, the redshift of galaxies, Cosmic Microwave Background, the abundance of deuterium, type II supernova data, is starting to come under the cross hairs as well for not supporting a literal six days creation. This appears to reflect a social abandonment of a world view that is supported by facts. This also currently has considerable political support. This does not bode well for our ability to grapple with the problems that face our world.

Much of the current debate involves the nature of truth, or what is thought of as truth. Science does not involve truth in a strict sense. Science deals with facts or datum, where these may support a particular theory or falsify it. A well established scientific theory supported by a huge body of evidence might be said to be tentatively true. Newton's law might be said to be tentatively true within some domain of observation and measurement, but has been shown in the past 100 years to be incomplete. Science is not able to prove a theory as true. The only subject that is able to prove anything true is mathematics. Religion on the other hand involves revealed truth, or an unprovable truth, that was revealed to certain prophets, which is accessible to believers through faith. Religious truth is believed, it is neither something proven mathematically, nor is it a theory supported by evidence. These revealed truths have been written in scriptural texts and form a part of the body of civilization. However, since science and religion involve different concepts of truth they are fundamentally different narratives. If this strict separation of science and religion is upheld then in principle there is no direct conflict. Yet there is a dynamic tension still apparent. For if one is to accept the big bang and biological evolution as the accepted model of reality it is not possible to believe a literal reading of Genesis. The reading of Genesis has to be taken according to questions of "why is there a universe and life," where science gives tentative truths about "how there is a universe and life."

This is similar to Stephen J. Gould's idea that science and religion are two separate magesteria, which is maybe just a temporary truce. However, it appears that this separation is disintegrating all around us. Outside of a few, most scientists are not involved with trying to disprove the Bible, or in showing God does not exist. On the other hand there is a huge, and socially popular, fundamentalist religious trend that is out to overthrow scientific understanding of the world. The revealed truths of religion are interpreted as facts that supercede theory supported by measurements and facts acquired through observation and experiment. This is seen particularly with the evolution vs. creation debate. This trend is not limited to Christianity, where similar trends are seen in Islam and ultra-orthodox Judaism. In the United States this trend has become highly politicized and its influence upon the health, education and welfare sector of government is growing. This trend is relatively alarming for a number of reasons, but a serious concern is that this is shifting the consciousness of society and its leadership away from a realistic world view required to address the problems confronting all of us this century.

That we are in this state of affairs has two principal and curiously opposite origins. The first was the liberal movement of the 1960s, where it became a sort of lore that one's own personal truth is no more or less than any other truth, even if that truth is scientific, which again is admitted to be a tentative truth. This "anything goes" attitude was seen in the explosion 35 years ago of various ideological and mystical trends. This is seen today in the Post Modernist trend, which says that all truth is completely relative, where one person's notion of truth is equivalent to anybody else's. The other source of this is due to Leo Strauss, who was the polar opposite of these liberal trends during the latter years of his life. Yet in a curious way his ideals come full circle to that of post modernism, but in an authoritarian manner. Strauss maintained his ideal for society as similar to Plato's *Republic*, which indicated that society should be run by a wise or philosophical elite. This ideal is similar to the notion of the "philosopher king," which Louis XIV fancied himself to be in the early 18th century. Strauss even went so far as to say that this society might have to be guided by various myths, even if those myths are known to be false. This is not that different from post modernism, but where those designated as the philosophical elite have the power to impose their myth. This is seen recently be comments by people in power, "We make our own reality." Various proteges of Strauss include powerful figures, such as Paul Wolfowitz, and his ideals form much of the canon behind the current ideological concepts that have plenary capacity.

The problem with the notion of completely relative truth is that it removes reason as a criterion for assessing whether some ideas are true, or tentatively true, or even plausible as distinct from ideas that are outright false or nonsensical. Ideas, whether they be a mathematical proof, or a good scientific theory, or somebody's mumbo jumbo, may be promoted on equal footing. To use the notion of memes, they are similar to genes that seek to promote themselves in the world. So this tendency for promotion will find avenues outside of reason. However, if the criterion of reason is weakened, then bogus ideas will promote themselves more successfully by other means. This often involves political power or religious belief. So the liberal Post Modernist environment is the perfect environment for the Straussian ideology, where the neo-elites in philosopher's robes promote their nonsense as the guiding myth for society. This is seen by the growing acceptance of Intelligent Design by political pressure, with no scientific endorsement.

So in the warf and woof of American society these two concepts have worked to create the current state of affairs. Religion has mushroomed in the US, where the revealed truth of the Bible is touted as at least equivalent to science, true to post modernist dictates. These religious truths are then set up as the ultimate truth, in accord with the Straussian notion of a guiding myth enforced by elites. This can be carried to other aspects of the socio-economic sphere in America. There is this growing cult of self-selected elites who are becoming increasingly arrogant and inward in cronism. The net effect of this is that society and its leadership are drifting away from a realistic view of the world. This is a disastrous trend. The United States is the leading power in the world, and its policies carry far beyond its borders. If this nation is trending down a path of self-created mythology it will drag much of the rest of the world along with it. This implies that the world is not well adapted to challenge the problems which confront it. Maybe the rest of the world will divorce itself from this path if this continues.

It is clear that for humanity to progress into the future, with all the technological, economic and social adaptations required, that the current trend seen in the United States must be ended. This end is not just an end to the political trajectory, but with the social trends as well. If humanity is to improve itself in the future it must do so realistically. If our species is to pursue its understanding of the universe the prevalent world view must be realistic, and not based upon myths upheld by self appointed elites who call themselves philosophers. The same is the case if the word philosopher is replaced by theologian or evangelical preacher. If things are to continue on increasingly mythic lines enforced by an elite it is far more likely that issues that confront the world will be managed in an increasingly incompetent manner. There is clear evidence of this already. Further, such a world is not one that is likely to invest much of its attention to understanding the world and universe in a realistic manner.

If these issues can be resolved in the forth coming decades humanity should be able to push the frontiers of knowledge further. This may include sending relativistic rockets or probes to the nearby stars. Concurrent with this will be further examinations of the universe, a real theory of quantum gravity and cosmology, an understanding of the origins of life, theoretical and empirical study on emergent complexity and self organization, possibly a science of consciousness, and an understanding of many other issues. If in the more distant future humanity is able to exist in some long term stable and sustainable manner interstellar probes might be sent to further stars, where there is some expectation that several centuries in the future people will be around to receive information from such probes. It may also be the case that other large scientific questions will require considerable time to answer, such as the origin of the universe and life and the nature of consciousness. In general the universe is open to us with questions, and the opportunity to address them is there, if we choose to do so.

The study of extraterrestrial life, called the field of exobiology, may illuminate very deep aspects of the foundations of reality. The subject of superstring theories indicate that the fundamental quantity in the world are not particles, but strings. These are fields parameterized along a cord that vibrates, just as a guitar string vibrates. The modes of these vibrations give the spectrum of elementary particles. This initially confounded physicist's sense of things, where the objection was that there were no things string-like in nature. Of course there exists a very stringy thing in nature all around us, called DNA. DNA contains information that is a code for the construction of polypeptides. Just as the quantum string or superstring has quantum information in its modes of vibration, DNA contains information as well. So here we have on the most fundamental scale $L = \sqrt{8\pi}L_p$ something stringy that "expresses" qualities about the world and on a much larger scale there exists another stringy thing that contains the fundamental information of life.

The analogue continues in a curious way. The superstring is studied according to how it is anchored to a membrane. Open strings, those with ends that do not reconnect into a loop, are anchored to a membrane. This is similar to the anchoring of a string to a musical instrument. The field content of the string determines that of the membrane, just as the vibration of a violin string is transduced by the rest of the instrument. DNA is similar, in that it couples with large folded polypeptides. The major ones are ribosomes that read the DNA to construct polypeptides and helicase molecules that split a DNA double string and replicase that duplicates it. There are other DNA binding proteins that block its expression. The stringy things in the biological world in a similar way couple with extended objects of dimension larger than one, where its information is transduced. To carry this further, the gravity field exists on closed strings, which have their analogue with plasmids or closed rings of DNA.

This is a curious "recherche" of similar structure on different scales. Similar things also exist in physics, such as isospin in weak interactions and in the model of nuclear structure of bound states of protons and neutrons. Recently supersymmetry, a symmetry that inter-changes bosons and fermions and thought to underly quantum field theory, as been found as an emergent symmetry in lower energy nuclear physics. The Higgs mechanism which breaks the electroweak interaction at the TeV range in energy is similar in structure to the Curie point for magnetization and the onset of superconductivity. There are other instances of structure of a similar nature appearing in different domains and scales of physics.

To carry this further we will digress into a discussion on quantum mechanics, though mostly informally. The appearance of something stringy in the quantum world and within the macroscopic world seems to suggest some sort of connection that we might want to consider. The finding of life on other worlds would give some added information for the following discussion and ensuing speculation.

There has been a debate in physics for nearly 80 years. Quantum states exist in a superposition of elementary states, called eigenstates, just as a musical note have contain harmonics. A quantum particle can exist in two states at the same time. The classic case is with the Einstein-Rosen-Poldolsky argument. Assume there exists a spin zero particle that decays into two particles of opposite spins, such as an electron and a positron. The spin state of the two particles are correlated to each other, and the wave function only indicates a probability for one of the particles to be in a certain spin orientation and the other in a counter orientation. Assume that the two particles move in opposite directions due to a Newtonian third law recoil after they are produced in a decay process. Further assume that one particle enters a region with a magnetic field, while the other is far removed. A spinning charged particle in a magnetic field will exhibit a precession, much as a gyroscope's angle of tilt will itself rotate. The particle subjected to the magnetic field will precess, so its spin representation according to some direction will be a spin vector that rotates around the lines of magnetic field. Because these two particles are in an entanglement, or entangled state, the other particle will also precess. This will be the case no matter how far apart the two are. Further, if one particle is measured to have a spin in a certain direction the other particle's spin is found to be in the opposite direction. This is the case even if the two measurements are performed simultaneously, or within a time frame shorter than it takes light to travel between the two detectors. By this entanglement the two particles sense each other's state, but they do so without communicating any information between each other. This is a nonlocal effect of quantum mechanics, which is something a minority of physicists strongly object to, and continue to this day. Further, this imposes a dichotomy between the quantum and classical domains of physics.

How a measurement occurs has been a source of controversy. The Copenhagen interpretation of quantum mechanics places a dichotomy between the quantum domain of reality and the classical domain, where the measurement apparatus is classical. For a quantum state in a superposition $|\psi\rangle = \sum_n C_n|n\rangle$ a measurement reduces the state to a particular eigenstate $|m\rangle$ corresponding to the eigenvalue of an observable the detector measured, O_m. This has been seen as some what artificial. If the detector is ultimately quantum mechanical the position of the measurement "needle" must also exist in some superposition of states. In other words the needle should be is various positions at once. This detector is put into an entangled state with the system it measures. Yet this does not appear to obtain in the real world.

Erwin Schrödinger advanced a thought experiment to illustrate a curious feature of this. An atomic nucleus in an excited state has a probability that it will decay over a certain period of time. Suppose that this nucleus is placed in a box with a detector. Since the detector is ultimately quantum mechanical the detector will enter into an entangled state, or a superposition, of decay detected and decay not detected. Now let us assume that the detector is attached to a device that opens up a pressurized vessel of poison gas. Because this is a detector of the detector the valve is then itself in a superposition of "open" and "closed" states. Then to complete this picture there is a cat in a closed cell with this set up. Again by the above this cat will be thrown into a superposition of being alive and dead. Now something is horribly amiss here. It simply can't be the case that a cat is quantum mechanically in a superposed state of alive and dead.

The problem is that things like detectors and cats, while composed of quantum mechanical entities such as atoms and electrons, are not in a pure quantum state. They are much larger than any quantum wavelength they would have by deBroglie's formula, and further have a temperature which signifies a measure of disorder that is uncharacteristic of a pure quantum state. This might be seen with hydrogen atoms. A single hydrogen atom has a quantum wave function. Given two hydrogen atoms their phases are random, or one is not determined by the other. So the two atomic wave functions interfere with each other. If we consider the phase as a vector on the argand (complex) plane, then the two phases determine a vector sum for the total phase. Now consider several atoms with phases given by vectors on the complex plane. In this case on average the vector sum is going to be pretty small, since the individual atomic phases will be randomly oriented. Now consider a mole of hydrogen-atoms. The total phase is going to be tiny, nearly zero, since they will all largely cancel each other out. A gas of hydrogen atoms in a bottle does behave in ways that are not very quantum mechanical. Using Poisson statistics the fluctuation involved with the total quantum phase for a mole of atoms would be $\sim 10^{-12}$ which is negligible. While the quantum superposition of the measured quantum system is lost its quantum phase is buried in a chaos of interfering phases so that its quantum influence on the apparatus is negligible. This gives a taste of the quantum/classical dichotomy.

The classical world is then due to some massive cancellation of quantum phases. This phase randomization and mixing is called decoherence, which seems to account for the nature of a hydrogen gas well enough. This sounds good, but it raises the question of why is the classical world what it is? In other words the Alive/Dead state of the cat appears to be still ultimately quantum mechanical. If this is the case the "alive" and "dead" states are stable against decoherence, while their superpositions are not. In other words we can look into the enclosed cage to see whether the cat is alive or dead, which seems to be due to a quantum effect. Yet is the living state of a cat really a proper quantum eigenstate?

So what is a cat? It is a 5–10 kg assemblage of biomolecules that exchanges material and energy with its environment, by eating, breathing, drinking, and excreting, where in every instance of this its quantum state, even as a mixture of quantum states, is being continually changed. In fact this happens every time a photon enters either of the cat's eye. The cat is in a state of quantum decoherence, and its continual mass-energy exchange with its environment perpetuates the inability to quantify what defines the quantum state of a cat. The problem is intractable. In order to have a cat as a quantum detector that enters into a superposed state with the quantum system measured the cat must also exist in some quantifiable quantum state. Yet it appears that this is not possible.

So it appears that the classical or macroscopic world involves states of existence that are not described by quantum mechanics. This tends to put us all back in the situation with the Copenhagen interpretation. And it has to mentioned that this interpretation has worked very well. Wojeck Zurek has advanced a theory of einselection to describe the classical world. These are states selected out by decoherence as the states of the classical world. The states of the classical world are then stable, but not so in large scale superpositions. While the theory appears reasonable in principle it still does not give us much of an understanding of why the classical world is what it is. It does not give much hint as to what the einselected states are.

Given the wave function $|\psi\rangle = \sum_n C_n|n\rangle$ the density matrix $\rho = |\psi\rangle\langle\psi|$ contains diagonal terms and off diagonal terms. The diagonal terms give the probabilities for a measurement of an eigenvalue corresponding to that eigenstate, and the off diagonal terms describe superpositions of states. A measurement destroys the off diagonal terms. The phases associated with these terms are lost to the environment. Quantum information has a Von

Neumann entropy interpretation

$$S = -k\rho \ln(\rho)\,, \tag{15.1}$$

where k is the Boltzmann constant, which changes during decoherence or loss of entanglement phase to the environment. Where does this entropy measure of lost information go? It goes into infuriatingly complex entanglements with the world outside the system, which are intractably too complex to predict or follow. This suggests that the emergent properties of the classical world are a way that this lost quantum information emerges in a different form. Of course this is physics at the frontiers of current research, and is in many ways speculative. Doubtless there are those with objections to this idea. Yet it appears these massive entanglements have a vast number of probable outcomes which are unpredictable by purely quantum means due to their intractable complexity. It is similar to chaos theory, where it is impossible to predict how the loss of information in a system might propagate into the future to change the system's state in the future.

Of course one might object that this analogue is with deterministic chaos, which is predictable on an "in principle basis," if one had a computer with an infinite floating point capacity. Yet from an operational viewpoint, if there is no way that the emergent complexity of the classical world can be predicted quantum mechanically then there are no quantum operators or states for these classical or macroscopic complexities. Again if something can't be observed then for all practical purposes, or in an operational sense, it does not exist. Of course those purists would object that there must still exist a live/dead quantum operator for Schrödinger's cat, but if it can't be observed or experimentally tested then is such a stance worth anything?

So we have a classical world that exists, at least to a good degree of approximation, which exhibits structure that appears outside the domain of quantum mechanics. For example, quantization of a chaotic system results in phase interferences that largely cancel out chaotic features, though there is what is called "scarring" that exists which is a remnant of chaos. The classical world appears to have features to it that don't exist in superpositions, such as quantum cats. Yet, the classical or macroscopic worlds do have structures that are similar to those of the quantum world. The analogue between the superstring and DNA has this appearance. The classical/quantum difference is that DNA is resistant to being in a superposition of states. In my work on octonionic field theory and extended quantum coding systems for quantum gravity, certain polyhedral representations exist for quantum states. This structure appears to underlie string theories. Yet what is interesting is that the next level up in the biological world from DNA is the virus. Many viruses have protein coats which are polyhedral in structure. It appears that what may be a structure underlying the string appears classically on a scale larger than DNA. Further, my work may well be extended to various higher order sporadic groups for the quantum coding structure, which may lead all the way up to the monster group. These sporadic groups have polyhedral representations with Coxeter groups. Whether this has analogues in the nature of life more complex than a virus is unknown, but it is interesting to ponder the possibility.

Life also emulates the quantum world in another way. The quantum world exhibits entanglements. Indeed, the classical world exists because entanglements are largely cancelled out. As such quantum mechanics does confront us with the fact that everything that exists does so by its relationship with everything else. Life also exhibits a form of entanglements in the web of life. While this is not at all a quantum entanglement, there are no EPR issues involved, all living things exist according to their relationship with every other living thing. This web of life can span across identified ecosystems, where webs of ecosystems form biomes and ultimately the entire biosphere of Earth. This appears to extend to relationships with the structure of the Earth, and as seen in the last chapter with the structure of the solar system. Both quantum mechanics and ecology challenge the classical notion of ontology where things are presumed to exist completely in of themselves. This thinking on the nature of existence is a coarse grained approximation. On a fundamental level everything that exists does so by its relationships with everything else.

If this recherche relationship between fundamental structures of the quantum universe and macroscopic structures, such as life, exists it would suggest that life is a repeated phenomenon throughout the universe. Life would under these circumstances be something that has a regular occurrence in the universe, which would suggest that biologically active planets have some density of occurrence amongst stars in the galaxy. What this density is remains unknown, but it should not be zero outside that of the Earth. This would indicate we might optically image a biologically active planet. If one is identified close enough to Earth we might be able to send a probe to it.

Relatively young G-class stars with terrestrial planets might prove to be of considerable interest as well. They may be planets where the geochemical processes which lead to life on Earth are taking place. It is known that nucleotides and amino acids can chemically emerge under some circumstances not that extra-ordinary. The Miller jar, due to Stanley Miller and Harold Urey, with reducing gasses of ordinary chemicals, such as ammonia, methane, hydrogen and carbon dioxide, will result in amino acids when subjected to electrical arcs. This has been referred to as the primal soup of organic materials. Of course the exact nature of the Earth's early atmosphere is not well known, but this is a relative approximation to what may have existed. Other mixtures have been tested, where there is a reduced production of amino acids if the simulated atmosphere is less reducing. However, once those compounds exist, which are monomers that compose more complex polymers in molecular biology, there must then have been some mechanism that permitted them to polymerize. It is also uncertain how extensive this soup would have been on the early Earth. It is unlikely to have been very extensive, for there should then be a geological layer of pyramidines and nitrogenous compounds that has not been found. However, polycyclic aromatic hydrocarbons (PAHs) have been found in nebula and on a meteorite that came from Mars. So it does appear that there are natural processes which form the basic building blocks of molecular biology.

It is likely that some energetic pathway for the beginning of biochemistry came from geological sources. Hot vents have metal sulfide bubbles that provide a barrier between temperatures, and are a source of energy through redox reactions. Some tripeptides have been found to occur in experimental simulations. It is then possible that oligomers and longer peptides will then be synthesized. It is likely that this was the source of energy for a complex mix of biochemicals that began to enter into self-replication. With this was a mix of RNA or DNA chains that interacted with polypeptides to duplicate each other. Once a measure of complexity was achieved the beginnings of evolutionary selection for biochemistry began to set in.

A planet around a relatively young G-class star, such as ϵ Eridani and β Comae Berenices, might hold a young rocky planet where pre-biotic processes are taking place. While this planet may not exhibit biological signature by optical means if it is in an appropriate orbit it could represent an Earth-like planet in its early justation. Such a planet could be an intellectual gold mine for our understanding of life in the universe. It is possible that Mars had features of this sorts as well, where evidence of this might exist in its geology and the chemistry of its geology. Mars may provide the first indications on how life started. A planet around a young star may well demonstrate this in real time.

This speculation can be taken to the extreme. If ever more complex structures in the macroscopic world reflect structures on a deeper fundamental level of physics and cosmology, then what about mental awareness and intelligence? This tends to suggest something bordering on the theological. It is my thinking that physical laws ultimately emerge by completely random means. This implies a fundamental vacuum structure that is highly random, in fact maximally random. Chaitan has found connections between Gödel's theorem and randomness. In effect the rules of mathematics exist as accidents of self-referential truth. Gödel'theorem involves mathematics that encodes or examines mathematics. There are some rather striking results from this. In particular a proposition "this proposition is unprovable" must be true, for the converse of this is a contradiction. It turns out that no mathematical system can avoid the existence of these odd propositions that state their own unprovability. It is possible that beneath physical symmetries or "laws" on the Planck scale $L_p \simeq 10^{-33}$ cm laws virtual states of a completely random quantum vacuum are self-referential, and the existence of the universe we observe may be one of these self-referential accidents. This has the consequence that not only might quantum fluctuations of spacetime generate cosmologies similar to ours, but that on a deeper level there might exist cosmologies which are organized according to principles utterly different than our own. Max Tegmark has speculated about this, where there might exist cosmologies that obey physics based on totally different mathematics, where most of these may involve mathematics we are ignorant of.

It may well be that consciousness involves self-reference, or at least some approximation of it that avoids some of the infinite aspects of it. If this is the case, then might this self-referential chaotic vacuum be something similar to consciousness? If this exists it might be interpreted as similar to what is commonly called God. Others might object to this interpretation, and the idea that a self-referential vacuum on the Planck scale is a conscious entity is a metaphysical extrapolation. This theological bit is a speculation that has gone off towards very extreme ends.

If the universe is a praedial system, or one that reflects its deepest foundations in the occurrence of viruses, bacteria, trees, nematodes, and people, then there should also be intelligent life elsewhere in the universe. Maybe we will find such intelligent life in the universe by their radio transmissions. This is Sagan's *Cosmic Connection*, where maybe over long periods of time intelligent life in at least local regions of the universe communicate with each other. If such communications can be established in the future with intelligent life in the universe can compare their various notebooks on how they have interpreted the script of the universe. We might call it cosmic peer review.

We live in a universe of immense proportions and perplexity. That we exist within it at all is a marvel. It is a huge question that challenges the best of our minds. It is unlikely that humanity will ever achieve the El Dorado of Voltaire's *Candide*, where all scientific answers were answered. Scientific and technological progress show no sign of abating by themselves. Even though there are social trends against the intellectual thrust of science, primarily from the religious front, it is curious to note that the Christian fundamentalists have enthusiastically embraced and used multi-media technology. Other problems such as energy and resource depletion and global ecological issues may blunt future science, yet it is also just as likely that science and technology will be the methods and tools by which we prevent these problems from becoming a disaster. After all it has not been either religious preachers nor luddites who have illuminated these problems, it was the scientists who did. The one thing that is also certain is that people want their world to stimulate their minds or at least their attention. The human brain is a sort of information feeding machine that does not like to have that flow cut back. In fact this is the thing that most prisoners find hard to adjust to, a world of complete routine and boredom. It is likely that even if the world does go into some static mode, either through some repressive system of control, a dark age due to resource or other limits that eventually people will figure a way out of it.

As indicated above everything exists ultimately because of its relationships with everything else. Everything that goes on with our world involves a communicating network and relationships between people. This is the case with science as well. The problems that we currently face ultimately come down to problems with our relationships with each other. A technology is nothing more than some highly configured set of materials and energy. The fundamental problem is not the technology, but rather our relationships within the context of that technology. This is the case whether the issue involves nuclear armaments, global warming, energy issues, or the media content on televisions. Ultimately the issue is far less about technology than it is the nature of our relationships with each other. By the same token the measure of future human progress will be far less measured by the advancements in technology, but by the content and depth of how we related to each other.

It is likely that the future will be one of vast change, even over the next few decades. Of course change is the thing that most politicos and the rest resist for the most part, and on both the so called liberal and conservative ends of the spectrum. They generally fail due to a lack of imagination, but instead have a plan that the world should "stick to." This is seen in science fiction programs and books as well. They generally depict a world organized in ways similar to today projected into some future scenario. *Star Trek* is a perfect example, a program which originated during the ramp up to the Apollo missions to the moon, and assumed a sort of space faring society organized around the type of engineering thought of the 1960's space program. The movie *2001 A Space Odyssey* portrayed a society that was interplanetary, and where the emotional affect of its characters were about as flat as the computer HAL 9000 that went awry. Of course people in today's world are not so flat in their personal expressiveness. Of course these movies were incapable of taking into consideration the ultimate chaos factor in any future, which are human beings.

Having said this, *2001 A Space Odyssey* is a movie that stated something few movies have said. When Commander Poole entered the Monolith, the secrets of the universe, unknown to him or anyone else, were made known to him. It is a bit revelationist, but many great thinkers have said their great leaps of thought have occurred this way. The mind is very strange.

Of course if human progress is to go forward people will become more aware of the developments of modern science, which include the bizarre nature of the quantum, general relativity and the breakthroughs in cosmology, which in recent years have been rather stellar. It is encouraging that books that popularize science have grown in their sophistication and from some notable sources. The simple fact is that the biblical statement that the truth will win out is something true. It just turns out that the truth here is something different than many biblical upholders believe.

A part of forging functional relationships between ourselves is some reconsideration of what we mean by truth. If in a century or so we are sending probes to other stars it can only be after we have made some sort of peace between our various notions of truth. There are three basic models of truth, mathematical, scientific and revelatory or religious. Mathematical truth is the firmest of them. A theorem that is proven true is absolutely true. Even though this involves relationships between abstract entities, a proof concerning some relationship between these entities, which can involve numbers or geometric forms or other things, is a hard truth. It does have to be admitted that Gödel's theorem has put some strains on this, and where further there are now various "proof theories" for various mathematical methods. There is then scientific truth, which is a contingent truth.

Science deals with facts, where these facts are concrete measurements and observations. These facts in total are then used to see if they support the predictions of a theory. If they do the theory is substantiated, but not proven. If the facts contradict the theory the theory is falsified. For a good theory, one which is substantiated by a preponderance of prior datum, the theory is then said to be falsified outside some domain of observation. This would be the case with Newton's laws, which is a very good theory within certain limits of low velocities and masses and for scales much larger than the atom. Newton's laws are in some sense contingently true. Other good theories not so falsified are likely also to fail at some point, such as general relativity and quantum mechanics in the bizarre domain of quantum gravity. Then finally there is revelatory truth, which is the most problematic for those in the scientific world. This is truth revealed to a prophet or some type of person who then speaks a revealed truth to the world. This revealed truth is most often given to the seer by some mystical process, or through the divine words of a God, or some such similar events. Some religions have a great day of God, where in the future the veil that separates the physical world from the religiously described mysterious world beyond is lifted and God, and God's truth is made plainly known.

Neither of these three models of truth is going to go away any time soon. As much as religious fundamentalists can't stand the idea of Darwin, his scientific truth will persist onward, though maybe outside the United States. As much as some hard line scientists with an atheistic bend can't abide the idea of God or religion, religion will be a factor in the future of humanity for some time, and maybe as long as we are around. The mathematicians are rather lucky at being largely removed from this fracas. At this time religious organizations are mounting a vigorous campaign to remove aspects of the scientific canon from the educational field. This intrusion is being steam rolled forwards within the current political power play that has unfolded in the United States. Yet largely this can only lead to not only bad science, but bad religion as well. The reason is that science and religion discuss completely different issues. Religious truth involves questions and answers to "why." Scientific truth involves questions and answers to "how and what." This dichotomy is up for dispute, but for the sake of a truce this suffices for the time. The two domains of truth involve separate categories of the human mind and heart. There is no point in trying to impose a system of thinking that involves the meaning of things over a system of thinking of a process of things, and visa vera.

Information is related to heat, which means that it is diffusive. As such information can't be easily contained. This information can involve of course a wide array of thing, from great discoveries to the nefarious machinations inside secret rooms of power. Because of this it will never be possible to quash scientific knowledge. Since questions are also information it will never be possible to eliminate them. So barring some utter catastrophe it is likely that scientific progress will continue, even if it does so through periods of difficulty. There is some reasonable prospect for a continuation of our expanding knowledge of the universe. This may in the next 100 years involve the launching of probes to other stars.

It is possible that all we will ever be able to examine closely with probes are planetary systems in a sphere of 50 to 100 light years. This is a tiny bubble in the Milky Way galaxy, which is just one galaxy out of billions. Exotic methods of propulsion are unlikely on physical grounds. Explorations further into interstellar space involve time frames that have few historical precedents. Nation states and civilizations generally have a lifetime of around 500 years. History indicates that cultural paradigms become outmoded and that civilizations become recycled and merged into new civilizations. This suggests that interstellar probes might occur in some more distant future after our current civilization has passed away and a new paradigm for civilization has emerged. So a probe meant to transmit a signal back some 1000 years after launch might have a different audience here on Earth, or none at all. However, within the 50 light year radius around the sun exist plenty of stars that could be explored, with about eight times that many out to 100 light years. So for now the interstellar frontier is vast and wide open, even if it is just a tiny bubble within a very vast universe.

Appendix

This appendix lists G-class stars, the masses of the stars, the orbital radius of their known planetary companion, the approximate muliplied rate of drift compared to the Earth (Drift Time), and the ratio of the Lyapunov exponent for a putative 1 AU planetary orbit relative to the Earth's orbit. The drift time is an approximate measure of how more rapidly the planetary orbital changes relative to that of the Earth. A negative sign indicates an inward drift.

Table A.1.. Comparison of orbital chaos of identified planets with Earth.

Star	Mass	Orbital Radius	Drift Time	Lyapunov Exponent
HD 83442	0.350000	0.038000		
HD 83443-b	0.160000	0.174000	−67.550835	4.906045
HD 16141	0.215000	0.350000	−41.751476	4.424930
HD 168746	0.240000	0.066000	−20.221100	3.700078
HD 46375	0.249000	0.041000	−19.180561	3.647271
HD 108147	0.340000	0.098000	−21.681307	3.769776
HD 75289	0.420000	0.046000	−19.382143	3.657722
51 Peg	0.470000	0.050000	−19.545704	3.666120
G4 V	0.480000	0.046000	−19.382143	3.657722
HD 6434	0.480000	0.150000	−24.415222	3.888495
HD 187123	0.520000	0.042000	−19.220625	3.649356
HD 209458	0.690000	0.045000	−19.341574	3.655627
ups And	4.610000	2.500000	−274.599670	6.308469
HD 192263	0.760000	0.150000	−24.415222	3.888495
ϵ Eridani	0.860000	3.300000	3.334593	1.904953
HD 38529	0.810000	0.129300	−23.268127	3.840384
HD 179949	0.840000	0.045000	−19.341574	3.655627
55 Cnc	0.840000	0.110000	5.000000	4.000000
			−1.781346	1.296306

Table A.1. (*Continued*).

Star	Mass	Orbital Radius	Drift Time	Lyapunov Exponent
HD 82943	0.880000	0.730000	1.630000	1.160000
			558.425720	7.018292
HD 121504	0.890000	0.320000	−38.148785	4.334700
HD 37124	1.040000	0.585000	−102.424133	5.322275
HD 130322	1.080000	0.088000	−21.208447	3.747731
ρ CrB	1.100000	0.230000	−29.752064	4.086141
HD 52265	1.130000	0.490000	−67.820068	4.910020
HD 177830	1.280000	1.000000	$\to \infty$	$\to \ln(\infty)$
HD 217107	1.280000	0.070000	−20.395420	3.708657
HD 210277	1.280000	1.097000	1874.800415	8.229314
HD 27442	1.430000	1.180000	544.444702	6.992901
16 CygB	1.500000	1.700000	35.999992	4.276732
HD 74156	1.560000	0.276000		
HD74156-b	7.500000	4.470000	−5.534779	2.406914
HD 134987	1.580000	0.780000	−364.462677	6.591571
HD 160691	1.970000	1.650000	41.751480	4.424930
HD 19994	2.000000	1.300000	196.000046	5.971273
HD 213240	3.700000	1.600000	48.999992	4.585002
Gliese 876	1.980000	0.210000		
Gliese 876	0.560000	0.130000	−123.241486	5.507303
HD 92788	3.800000	0.940000	−4899.999023	9.190217
HD 8574	2.230000	0.760000	−306.249939	6.417555
HR810	2.260000	0.925000	−3136.000732	8.743987
47 Uma	2.410000	2.100000	14.578513	3.373088
HD 12661	2.830000	0.789000	−396.217377	6.675133
HD 169830	2.960000	0.823000	−563.056641	7.026499
14 Her	3.300000	2.500000	7.839999	2.753741
GJ 3021	3.310000	0.490000	−67.820068	4.910020
HD 80606	3.410000	0.439000	−56.049641	4.719411
HD 195019	3.430000	0.140000	−23.850727	3.865109
Gl 86	4.000000	0.110000	−22.269913	3.796552
τ Boo	3.870000	0.046200	−19.390272	3.658141
HD 50554	4.900000	2.380000	9.262757	2.920119
HD 190228	4.990000	2.310000	10.279121	3.024050
HD 168443	7.200000	0.290000		
HD168442-b	17.100000	2.870000	−9.689448	2.965071
HD 222582	5.400000	1.350000	143.999969	5.662964
HD 28185	5.600000	1.000000	$\to \infty$	$\to \ln(\infty)$
HD 178911	6.470000	0.326000	−38.831017	4.352423
HD 10697	6.590000	2.000000	17.639997	3.563584
70 Vir	6.600000	0.430000	−54.293625	4.687583
HD 106252	6.810000	2.610000	6.805293	2.612645
HD 89744	7.200000	0.880000	−1224.999756	7.803791
HD141937	9.700000	1.490000	73.469376	4.990030
HD68988	1.900000	0.070000	−21.395420	3.756506
HD142	1.360000	0.980000	11.388519	11.388519
HD4203	1.640000	1.090000	1.135507	8.379674
HD114783	0.990000	1.200000	440.999756	6.782212
HD23079	2.760000	1.480000	76.562485	5.031270
HD4208	0.810000	1.690000	37.051029	4.305503
HD33636	7.710000	2.620000	7.721537	2.738557
HD39091	9.940000	3.500000	3.822400	2.039706

Bibliography

[2.1] I. Newton, *Philosophiae Naturalis Principia Mathematica*, London 1687. Transl. by A. Motte (1729), and F. Cajori: *Sir Isaac Newton's Mathematical Principles of Natural Philosophy and his System of the World*, University of California Press, Berkeley 1934.

[2.2] R. S. Westfall, *Never at Rest, A Biography of Isaac Newton*, Cambridge U. Press (1980).

[2.3] C. Boyer, *A History of Mathematics: Henri Poincaré*, John Wiley & Sons 1968.

[3.1] W. T. Thomson, *Introduction to Space Dynamics*, Dover Publications (1986).

[3.2] G. P. Sutton, O. Biblarz, *Rocket Propulsion Elements*, John Wiley (2000).

[3.3] `http://www.grc.nasa.gov/WWW/ion/present/hipep.htm`.

[3.4] A. Bond, A. R. Martin, R. A. Buckland, T. J. Grant, A. T. Lawton, et al., Project Daedalus, *Journal of the British Interplanetary Society*, **31** (Supplement, 1978).

[4.1] A. Goetzberger, J. Knobloch, B. Voss, *Crystalline Silicon Solar Cells: Technology and Systems Applications*, John Wiley & Sons (1998).

[4.2] `http://www.ne.doe.gov/space/space-desc.html`.

[4.3] `http://saturn1.jpl.nasa.gov/spacecraft/safety/power.pdf`.

[4.4] Kenneth S. Krane, *Introductory Nuclear Physics*, Wiley & Sons (1988).

[5.1] R. R. Bate, *Fundamentals of Astrodynamics*, Dover Publications (1971).

[6.1] J. D. Jackson, *Classical Electrodynamics*, Wiley, 3rd edn. (1998).

[6.2] W. Rindler, *Essential Relativity: Special, General, and Cosmological*, Springer-Verlag, Rev/2nd edition (1980).

[6.3] K. S. Thorne, C. W. Misner, J. A. Wheeler, *Gravitation*, W. H. Freeman (1973).

[6.4] S. W. Hawking, G. F. R. Ellis, *The Large Scale Structure of Space-Time* (Cambridge Monographs on Mathematical Physics), Cambridge University Press (1973).

[7.1] C. Cohen-Tannoudji, Bernard Diu, Frank Laloe, *Quantum Mechanics*, Wiley-Interscience (1992).

[7.2] A. Zee, *Quantum Field Theory in a Nutshell*, Princeton University Press (2003).

[7.3] K. S. Thorne, C. W. Misner, J. A. Wheeler, *Gravitation* W. H. Freeman (1973).

[8.1] R. L. Forward, *Journal of Spacecraft and Rockets*, 1985, 0022-4650 **22**, no. 3, 345–350 (1985).

[8.2] M. Born, E. Wolf, *Principles of Optics: Electromagnetic Theory of Propagation, Interference and Diffraction of Light*, Cambridge University Press, 7th edition (1999).

[8.3] W. W. Mendell, Strategic Considerations for Cislunar Space Infrastructure, `http://ares.jsc.nasa.gov/HumanExplore/Exploration/EXLibrary/DOCS/EIC042.HTML`.

[8.4] R. L. Forward, Roundtrip Interstellar Travel Using Laser-Pushed Lightsails *Journal of Spacecraft*, **21**, 2 (1984).

[9.1] `http://www.nasa.gov/mission_pages/exploration/main/index.html`.

[9.2] A. Zee, *Quantum Field Theory in a Nutshell*, Princeton University Press (2003).

[9.3] K. S. Thorne, C. W. Misner, J. A. Wheeler, *Gravitation*, W. H. Freeman (1973).

[9.4] L. Crowell, *Found. Phys. Lett.*, **12**, 6, 585 (1999).

[9.5] L. Crowell, *Quantum Fluctuations of Spacetime*, World Scientific (2005).

[10.1] `http://en.wikipedia.org/wiki/Von_Neumann_probe`.

[10.2] `www.nano.gov`.

[10.3] W. E. Billups, M. A. Ciufolini (Editors), *Buckminsterfullerenes*, Wiley (1993).

[11.1] K. S. Thorne, C. W. Misner, J. A. Wheeler, *Gravitation* W. H. Freeman (1973).

[11.2] K. S. Thorne, *Black Holes and Time Warps: Einstein's Outrageous Legacy*, W. W. Norton & Company; Reprint edition (January 1, 1995).

[11.3] M. Alcubierre, The Warp Drive: Hyper-fast Travel within General Relativity, *Class Quantum Grav.* **11**, L73–L77.

[11.4] S. V Krasnikov, *Hyper-fast Interstellar Travel In General Relativity*, `http://arxiv.org/abs/gr-qc/9511068`.

[11.5] J. Magueijo, L. Smolin, Lorentz Invariance with an Invariant Energy Scale, *Phys. Rev. Lett.* **88** (2002).

[12.1] T. L. Kepner, *Extrasolar Planets: A Catalog of All Discoveries in Other Star Systems* McFarland & Company (2005).

[12.2] `http://www.princeton.edu/~willman/planetary_systems/Gliese581.html`

[12.3] B. Gladman et al., Synchronous Locking of Tidally Evolving Satellites. *Icarus* 122: **166** (1996).

[13.1] Claude Semay and Bernard Silvestre-Brac, The equation of motion of an interstellar Bussard ramjet, *Eur. J. Phys.* **26**, 75–83 (2005).

[13.2] Y. Kondo (ed.), *Generational Space Ships : Apogee Books Space Series 34*, Collector's Guide Publishing Inc. (2003).

[14.1] Korycansky, Laughlin, and Adams, *Astrophys. Space Sci.* **275**, 4 (2001).

[14.2] T. J. Crowley, *Journal of Climate*, **3**, pp. 1282–1292 (1990).
[14.3] R. A. Berner, *Science*, **249**, pp. 1382–1386 (1990).
[14.4] P. S. Laplace, The report in Parisian AS, 1773.
[14.5] J. L. Lagrange, The article 1784, in V. G. Djomin, *Destiny of a solar system*, Science, M. (1969).
[14.6] A. Poincaré, *Les Methods Nouvelles de la Mechanique Celeste*, Gauthier-Villars, Paris.
[14.7] A. M. Lyapunov, *Ann. Math. Studies* **17**, Princeton (1947).
[14.8] V. I. Arnold, The solution of a problem on motion stability in arbitrary Hamiltonian systems if there is permanent disturbances, in V. G. Djomin, *Destiny of a solar system*, Science, M. (1969).
[14.9] E. N. Lorenz, *J. Atmos. Sci.* **20**, 130 (1963).
[14.10] E. Ising, *Z. Physik* **31**, 253 (1956).
[14.11] G. Gonzalez, G. Wallerstein and S. Saar, *ApJ. Letter*, **511** (1999), `http://www.obspm.fr/encycl/encycl.html`.
[14.12] G. Gatewood, *BAAS* **511** (1999).
[14.13] G. Gonzalez, D. Brownlee and P. Ward, *Scientfic American*, **285**, 4 (2001).

Index

acceleration, 7, 27, 29, 55, 79
  centripetal, 12
  photon sail, 83
  proper time, 64
  relativistic, 59, 60
  relativistic rocket, 70, 72
Alcubierre warp drive, 113
alien, 138
angular velocity, 12
antimatter, 19, 67, 74, 75, 77, 94
Apollo program, 1, 18, 36, 129
appendix, 183
astrobiology, 42, 119, 125, 128, 163, 169
  G-class star, 121
astronavigation, 43
atomic bomb, 25, 37

baryon conservation, 78
baryon number, 78
Bayes rule, 157
binding curve of energy, 39
black hole, 78, 96, 98
  Kerr solution, 107
  Schwarzschild solution, 96
buckminsterfullerenes, 101
Bussard ramjet, 136

Carl Sagan, 2
Cauchy horizon, 110
centripetal force, 12
chaotic dynamics, 141, 149
  Ising model, 152
  iterated map, 151
  KAM theory, 144
  Lyapunov exponent, 154
  Poincaré section, 149
  resonance conditions, 150
chemical rocket, 5
coordinate time, 55, 69
curvature of spacetime, 65, 96

Daedalus, 25, 29, 41, 67
DNA, 101, 164, 176
  comparison with string theory, 170
Doppler method for finding planets, 128

Einstein field equation, 116
Einstein field equations, 114
Einstein, Albert, 49
electromagnetism, 27, 31, 52, 95
escape velocity, 43
event horizon, 97, 107
extrasolar system, 119, 154, 163
  Lyapunov exponent, 157
  Earth-like planet, 158, 159
  Lyapunov exponent, 160
  planet, 120, 125
extrasolar systems, 141

force, 7, 13, 27
  gravity, 13
  thrust, 18

general relativity, 64
Newton's second law, 82
relativistic, 59, 61
force, Newton's second law, 43, 44
frame, 49, 51, 52, 54, 57, 61, 69
inertial, 52
inertial and freely falling, 64
inertial and Newton's first law, 6
Fresnel lens, 84, 91
fusion, 25, 39, 40
ITER, 41
future of the Earth, 146

G-class star, 121, 124, 137, 159, 176
and F and K class stars, 123
gamma factor, 54, 69, 70, 76, 77
low gamma spacecraft, 71
Gliese 581, 125
Goddard, Robert, 5, 20
gravitation
Newtonian, 13
general relativity, 62–65, 112, 116
Newtonian, 8, 13, 63, 143, 144

harmonic oscillator, 143
Hertzsprung-Russell diagram, 121
Hohmann transfer, 46

inertial reference frame, 6, 10, 11
Newton's first law, 6
interstellar ark, 137
invariance, 144
invariant
interval, 55, 56, 59
ion rocket, 21, 22

Jupiter, 1, 22, 146, 149
interaction with Earth, 144
resonance with Earth, 150

Kepler's laws, 13, 45, 121, 144, 151
kinetic energy, 39, 43, 76, 79, 103
Krasnikov tube, 115

law of motion
relativistic, 59, 65
laws of motion, 6, 8, 11, 12
Newton, 104
relativistic, 71, 82
length contraction, 54, 73
light cone, 56, 96
Lorentz force, 27, 104
Lorentz transformation, 58, 59
Lorentz and Poincarë groups, 62

main sequence star, 142, 146, 154
main sequence stars, 121
Michelson-Morely experiment, 51
momentum, 103
Newtonian, 7, 10, 14, 17
photon, 75
relativistic, 28, 55, 58, 65, 69, 75, 77, 79
mometum
relativistic, 64

nano-bots, 102
nano-space probes, 101
nearby stars, 119, 122
Newtonian mechanics, 6, 8, 44
De Mundi Systemate, 6
Philosophiae Naturalis Principia Mathematica, 6
De Motu Corporum, 9
Nichols radiometer, 29, 79
nuclear fission, 38
nuclear physics
fission, 38
fusion, 40, 41
radioactive decay, 38
nuclear power
ITER tokamak, 41
NERVA, 23
nuclear rocket, 23
radioactive decay, 22, 31
reactor, 23
SNAP 10A, 37
nuclear rocket
Daedalus, 25
NERVA, 22
Orion, 25

orbital 1/f behavior, 149
orbital parameters, 45

pendulum, 11
photoelectric cell, 34
photoelectric effect, 33
photon pressure, 29, 79, 105
photon sail, 79, 91, 122
photovoltaic cell, 19
Planck length, 99
plasma physics, 26, 27, 40, 76
plutonium, 36, 38
Poincaré section, 149, 152
Poioncarë, 14
proper time, 55, 59, 61, 64, 69, 71, 110, 115
proton decay, 95

quantum black hole, 98
  Planck scale, 99
quantum black holes, 71
quantum mechanics, 3, 10, 15, 33, 40, 49, 66, 67, 75, 95, 170, 171
  and gravity, 66
  gravity, 177
  information, 174
  Schrödinger's cat, 173
quark, 95
quarks, 78

railgun, 103
  nano-railgun, 104
red dwarf star, 119, 125
relativistic rocket, 3, 55, 91, 136
relativity, 10
  general, 11, 62, 64, 65, 116
  special, 49, 55, 59, 62, 79, 82
Ricci curvature, 65
rocket
  chemical, 1, 5, 17, 18
  fusion, 25, 41
  nuclear, 22, 24, 25
  relativistic, 67, 91
  rocket equation, 17
    relativistic, 70

simultaneity, 57
solar sail, 28, 29
solar system, 6, 11
  stability, 8, 9, 15, 146
space colony, 129, 132, 135
specific impulse, 18, 19, 21, 23, 24, 26, 41, 70
speed of light, 1, 19, 29, 41, 50
  light cone, 56
  relativity, 54
Star Trek, 139, 165, 179
Star Wars, 1
starsail, 91
StarTrek, 1, 107
Starwisp, 79, 86, 88
string theory, 170

terrestrial planet, 126
three body problem, 9, 143
  first integral, 48
  Galois theory, 48
thrust, 18
tidal force, 126
tidal locking, 127
time dilation, 54, 73, 81, 110
Tsiolkovsky, 5

uranium, 24, 38

VASIMR, 19, 26, 39
vector, 13, 15, 54, 58
  Killing, 64
  magnetic, 27
  parallel transport, 65
von Braun, Werner, 5
von Neumann probe, 101

warp drive, 113
wormhole, 109